Worked Examples
in
Engineering Mathematics

Worked Examples
in
Engineering Mathematics

L. R. Mustoe
Department of Engineering Mathematics,
Loughborough University of Technology

JOHN WILEY & SONS

Chichester · New York · Brisbane · Toronto · Singapore

Library of Congress Cataloging-in-Publication Data:

Mustoe, L. R.
  Worked examples in engineering mathematics.
  Includes index.
  1. Engineering mathematics—Problems, exercises, etc.
I. Title.
TA333.M87   1986   620′.001′51   86-13305
ISBN 0 471 91171 2

British Library Cataloguing in Publication Data:

Mustoe, L. R.
  Worked examples in engineering mathematics.
  1. Engineering mathematics—Problems,
  exercises, etc.
  I. Title
  510′.2462   TA333
  ISBN 0 471 91171 2

Printed and bound in Great Britain

To my mother
  and to Professor Avi Bajpai, OBE,
  my mentor and friend

Contents

Preface

The author's experience in teaching undergraduates and in marking examination scripts, both as an internal and an external examiner, has led to the conviction that most students lack the ability to apply their mathematical knowledge to non-standard problems. Whereas they may be capable of performing simple exercises in manipulation they fall down badly when required to use the same skills in a different or unfamiliar setting. 'If only I could see how to start' is an all too common remark made by students.

It is to help to overcome this weakness that this book has been prepared. The expectation is that the careful study of the problems and their solutions will be a complement to the student's lecture notes and/or the recommended conventional textbook. Not only should the student's skills in problem-solving be enhanced, but his understanding of the mathematical techniques employed should be increased.

The solutions are not intended to be model answers, but outlines with extra detail added when it is felt advisable to help readers over difficult parts. Where additional explanation is given in a solution over and above that necessary to answer the given problem, this is indicated by a vertical line in the left-hand margin. Because of the particular weakness of students in the sketching of curves, a summary of the main points to be borne in mind is given before the problems on that topic.

A collection of basic results which may prove helpful is provided at the end of this book.

Many of the examples are taken from the Part I examinations of the Council of Engineering Institutions (CEI), now The Engineering Council (EC). The majority of the remainder are from examination papers set by the author for Loughborough University of Technology

(LUT). Consequently, this book is suitable for candidates for the Engineering Council Part I examinations as well as for students of engineering at a level indicated by Bajpai, Mustoe, and Walker, *Engineering Mathematics*, John Wiley (1974). Indeed, the problems which follow have been used by the author in teaching courses based upon the earlier book.

Thanks are due to The Engineering Council and to Loughborough University of Technology for permission to use questions from their past examinations. The solutions are the author's own and not those of the Council or of the University.

The author is grateful to those students who have given helpful comments on the problems and proposed solutions. He accepts full responsibility for any errors or inaccuracies that may be found.

It is a pleasure to acknowledge the help provided by the staff at John Wiley and Sons.

Finally, one debt of gratitude is gladly acknowledged: it is impossible to over-estimate the benefit that the author has gained from the encouragement and inspiration given to him by Professor A. C. Bajpai.

A

Complex Numbers

Example 1

(i) The centre of an equilateral triangle is represented by $-1 + i$ and one vertex by $3 + 5i$. Find the complex numbers representing the other vertices.

(ii) The complex number w is related to the complex number z via $w = (z - 1)/(z + i)$.

    (a) If w is purely imaginary show that the locus of z is a circle; find its centre and radius.

    (b) If w is real show that the locus of z is a straight line; find its intercepts with the axes.

(i) The key here is to translate the origin to the point $(-1, 1)$ via $w = z + 1 - i$. The centre of the triangle, E, goes to the new origin $0'$ and the given vertex A to the point $A' = (4, 4)$. See Figure 1. A' can be written as $4\sqrt{2} \angle 45°$.

By the properties of an equilateral triangle, angle $B'0'A' = $ angle $A'0'C' = 120°$ and therefore B' is $4\sqrt{2} \angle 165°$ and C' is $4\sqrt{2} \angle 285°$. Expressing B' and C' in cartesian coordinates $B' = -5.464 + 1.464i$ and $C' = 1.464 - 5.464i$; then subtracting $1 - i$ to revert to the original system we obtain

$$B = -6.464 + 2.464i, \qquad C = 0.464 - 4.464i$$

(ii) $\quad w = \dfrac{(x - 1) + iy}{x + i(y + 1)} = \dfrac{x - i(y + 1)}{x - i(y + 1)} \cdot \dfrac{(x - 1) + iy}{x + i(y + 1)}$

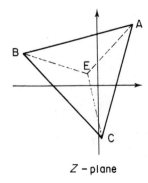 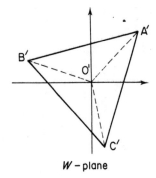

Z – plane   W – plane

Figure 1

(a) The real part of w is

$$\frac{x(x-1) + y(y+1)}{x^2 + (y+1)^2} = 0$$

Hence $x(x-1) + y(y+1) = 0$ or

$$\left(x - \frac{1}{2}\right)^2 + \left(y + \frac{1}{2}\right)^2 = \left(\frac{1}{\sqrt{2}}\right)^2$$

This is a circle, centre $x = \frac{1}{2}$, $y = -\frac{1}{2}$ and of radius $\dfrac{1}{\sqrt{2}}$

(b) The imaginary part of w being zero means that

$$yx - (y+1)(x-1) = 0$$

i.e. $y - x + 1 = 0$. This represents a straight line and both $(0, -1)$ and $(1, 0)$ lie on the line.

Example 2

Find the region in the Argand plane defined by

(i) $|z-1|/|z-2| < 1$ (ii) $0 < \arg[(2-2)/(z-1)] < \pi/2$

(iii) $|z-1| + |z-i| < 4$.   (CEI)

Geometrical approaches are often neater and simpler than analytical ones in the solution of domain description problems.

(i) *Algebraically*

$$|z - 1| < |z - 2|$$

therefore

$$|z - 1|^2 < |z - 2|^2$$

i.e. $(x + iy - 1)(x - iy - 1) < (x + iy - 2)(x - iy - 2)$

i.e. $(x - 1)^2 + y^2 < (x - 2)^2 + y^2$

i.e. $-2x + 1 < -4x + 4$

i.e. $x < \dfrac{3}{2}$

Geometrically $|z - 1|$ is the distance of the point z from the point 1 therefore we look for all points which are closer to 1 than 2. Refer to Figure 2(a).

The line $x = 3/2$ contains all points equidistant from 1 and 2; we seek the half-plane to its left, i.e. $x < 3/2$.

(ii) In Figure 2(b), the line segment AP represents $(z - 1)$ and $\hat{BAP}$ is $\arg(z - 1)$. Similarly BP represents $(z - 2)$. Then

$$\hat{APB} = \arg(z - 2) - \arg(z - 1) = \arg\left[\frac{z - 2}{z - 1}\right].$$

If AB is a diameter of a circle on which P lies then $\hat{APB} = \pi/2$. Above the real axis P lies outside the circle when $0 < \alpha < (\pi/2)$. What about the region below the real axis?

Convince yourself that there is no part of the domain there.

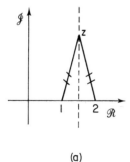

(a)

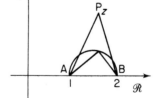

(b)

Figure 2

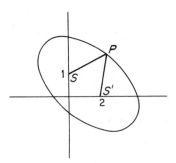

Figure 3

(iii) A well known property of an ellipse is that the sum of the distances to the foci from a point on the perimeter is constant. Hence the perimeter of the ellipse with points $(1,0)$ and $(0,1)$ as foci is given by $|z-1| + |z-i| = $ const. See Figure 3. The larger the value of the constant, the larger the ellipse. If the equals sign is replaced by a less than sign the region described is the interior of the ellipse. (Simply check one of the foci against the inequality.)

The cartesian equation of the ellipse can be shown to be

$$60x^2 + 60y^2 - 8xy - 64x - 64y - 192 = 0$$

However, to recognize this as the equation of an ellipse and then identify the foci is not easy. We merely note in passing that the terms $64x$ and $64y$ relate to the fact that the centre of the ellipse is not at the origin and the term $8xy$ is the result of the major and minor axes of the ellipse not being on the x and y axes respectively.

Example 3

If $z = x + iy$ and $w = u + iv$, where $x, y, u,$ and v are real and if $z = 2i/w$, express x and y in terms of u and v.

Suppose that the point (x, y) moves round the trapezium ABCD with vertices A(1, 1), B(1, −1), C(2, −2), and D(2, 2). Find the equations of the path traced by the point (u, v) and the coordinates of the points corresponding to A, B, C and D. Sketch the resulting path.

Suppose further that G(−1, −1) and H(−1, 1) are two other points in the *z*-plane. What area in the *w*-plane corresponds to the inside of the square ABGH? (CEI)

$$x + iy = z = \frac{2i}{u + iv} = \frac{2i}{u + iv} \cdot \frac{(u - iv)}{(u - iv)} = \frac{2v}{u^2 + v^2} + i \cdot \frac{2u}{u^2 + v^2}$$

Hence

$$x = \frac{2v}{u^2 + v^2}, \qquad y = \frac{2u}{u^2 + v^2}.$$

Refer to Figure 4.

$$AB: x = 1 \rightarrow 2v = u^2 + v^2 \rightarrow u^2 + (v - 1)^2 = 1$$
$$BC: x + y = 0 \rightarrow 2v + 2u = 0 \rightarrow v + u = 0$$
$$CD: x = 2 \rightarrow v = u^2 + v^2 \rightarrow u^2 + (v - \tfrac{1}{2})^2 = \tfrac{1}{4}$$
$$DA: y = x \rightarrow v = u$$

Let A in the *z*-plane correspond to A′ in the *w*-plane etc. Then

$$A \rightarrow A' = (1, 1)$$
$$B \rightarrow B' = (-1, 1)$$
$$C \rightarrow C' = (-\tfrac{1}{2}, \tfrac{1}{2})$$
$$D \rightarrow D' = (\tfrac{1}{2}, \tfrac{1}{2})$$

Note that $E = (1, 0) \rightarrow E' = (0, 2)$ and $F = (2, 0) \rightarrow F' = (0, 1)$

The resulting path A′B′C′D′ is shown in Figure 5.
Now

$$BG: y = -1 \rightarrow (u + 1)^2 + v^2 = 1$$
$$GH: x = -1 \rightarrow u^2 + (v + 1)^2 = 1$$
$$HA: y = 1 \rightarrow (u - 1)^2 + v^2 = 1$$

$$G \rightarrow G' = (-1, -1)$$
$$H \rightarrow H' = (1, -1)$$

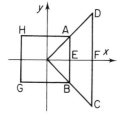

Figure 4

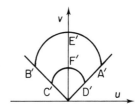

Figure 5

6

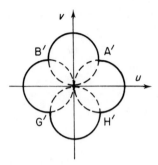

Figure 6

The area *inside* ABGH becomes the area *outside* the union of the four circles shown in Figure 6. For example, the point $(\frac{1}{2}, 0)$ in the z-plane leads to

$$\frac{1}{2} = \frac{2v}{u^2 + v^2}, \qquad 0 = \frac{2u}{u^2 + v^2}$$

whereupon $u = 0$ and therefore $2v/v^2 = \frac{1}{2}$ so that $v = 4$. The corresponding point in the w-plane is $(u, v) = (0, 4)$, which lies outside the boundary A'B'G'H'.

Example 4

(i) Solve the equation $\sin z = 2$.
(ii) If $|z| = a$ and $Z = z\,e^{i\theta}$, what is the locus of Z as z moves around the curve above? If also $W = \frac{1}{2}[z + (a^2/z)]$ show that W will describe a straight line segment.
(iii) The current entering a telephone line is the real part of

$$\frac{\cos \omega t + i \sin \omega t}{\cosh(s + is)}$$

Express this in the form $A \sin(\omega t + \alpha)$.   (CEI)

(i) Since $e^{iz} = \cos z + i \sin z$, $e^{-iz} = \cos z - i \sin z$ and therefore $e^{iz} - e^{-iz} = 2i \sin z$. The equation $\sin z = 2$ becomes

$$e^{iz} - e^{-iz} = 4i$$

or

$$(e^{iz})^2 - 4i(e^{iz}) - 1 = 0$$

Hence

$$e^{iz} = (2 + \sqrt{3})i \text{ or } (2 - \sqrt{3})i$$

Now

$$\log w = \log |w| + i(\arg w + 2k\pi), k = 0, 1, 2, \ldots$$

so that

$$\left.\begin{array}{l} iz = \log_e(2 + \sqrt{3}) + i(\pi/2 + 2k\pi) \\ = \log_e(2 - \sqrt{3}) + i(\pi/2 + 2k\pi) \end{array}\right\} k = 0, 1, 2, \ldots$$
or

Finally

$$\underline{z = (\pi/2 + 2k\pi) - i\log_e(2 \pm \sqrt{3})}$$

(ii) If $z = r\,e^{i\theta}$ then $Z = z\,e^{i\theta} = r\,e^{2i\theta}$ so that as θ takes values 0 to 2π and $r = a$ then Z travels the same locus as z but twice round the circle $|z| = a$.

$W = \frac{1}{2}(a\,e^{i\theta} + a\,e^{-i\theta}) = a\cos\theta$. When θ goes from 0 to 2π, $\cos\theta$ goes from 1 to -1 and back again. Note that W is a real number, in fact, the real part of z.

As z moves once round the circle anti-clockwise from B, W moves along the diameter from O to B, B to O, O to A and A to B. See Figure 7.

(iii) $\cosh(s + is) = \frac{1}{2}(e^{s+is} + e^{-s-is})$
$= \frac{1}{2}e^s[\cos s + i\sin s] + e^{-s}[\cos s - i\sin s])$
$= \cos s.\cosh s + i\sin s.\sinh s$

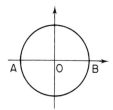

Figure 7

Then

$$\frac{\cos \omega t + i \sin \omega t}{\cosh(s + is)} = \frac{\cos \omega t + i \sin \omega t}{\cos s \cosh s + i \sin s \sinh s}$$

$$\times \frac{(\cos s \cosh s - i \sin s \sinh s)}{(\cos s \cosh s - i \sin s \sinh s)}$$

This has real part

$$\frac{\cos \omega t . \cos s . \cosh s + \sin \omega t . \sin s . \sinh s}{\cos^2 s . \cosh^2 s + \sin^2 s . \sinh^2 s}$$

Now

$$A \sin(\omega t + \alpha) = A \cos \omega t . \sin \alpha + A \sin \omega t . \cos \alpha$$

Comparing the two expressions,

$$\frac{A \cos \alpha}{A \sin \alpha} = \cot \alpha = \frac{\sin s . \sinh s}{\cos s . \cosh s} = \tan s . \tanh s$$

and

$$A^2 \sin^2 s + A^2 \cos^2 s = A^2 = \frac{\cos^2 s \cosh^2 s + \sin^2 s \sinh^2 s}{(\cos^2 s \cosh^2 s + \sin^2 s \sinh^2 s)^2}$$

$$= \frac{1}{\cosh^2 s \cos^2 s + \sinh^2 s \sin^2 s}$$

$$= \frac{1}{\cosh^2 s (1 - \sin^2 s) + \sin^2 s \sinh^2 s}$$

$$= \frac{1}{\cosh^2 s - \sin^2 s}$$

Hence we can find α and A.

Example 5

(i) Show on an Argand diagram the points A($\exp 3i\theta$) and B($\exp i\theta$). Hence if O is origin construct the parallelogram OAPB with OA parallel to BP and OB parallel to AP. Write the complex number z represented by P in both cartesian and exponential form.

(ii) Indicate further the complex points $\bar{z}$ (complex conjugate) and

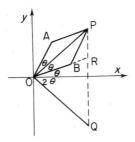

Figure 8

$z^{1/2}$. Show under what conditions $z^{1/2}$ lies on OB and between O and B, for $0 < \theta < \pi/2$.

(iii) Let A be fixed with argument 3α and let the argument of B vary. Sketch the curve described by P and find OP as a function of θ.

(CEI)

(i) $$OA = OB = 1$$

The segments OA and BP both represent $e^{3i\theta}$ in the sense that the length of OA = length of BP = $|e^{3i\theta}|$ = 1 and the angle made by either segment with the real axis is the argument of $e^{3i\theta}$, namely 3θ. P is given by $z = e^{3i\theta} + e^{i\theta}$ (parallelogram law of addition). That is $z = e^{2i\theta}(e^{i\theta} + e^{-i\theta})$ or $r e^{i\alpha} = z = 2 \cos \theta$. $e^{2i\theta}$ so that $r = 2 \cos \theta$ and $\alpha = 2\theta$. Hence $x = r \cos \alpha = 2 \cos \theta \cos 2\theta$ and $y = r \sin \alpha = 2 \cos \theta \sin 2\theta$ or $x + iy = z = 2 \cos \theta . \cos 2\theta + i2 \cos \theta . \sin 2\theta$.

(ii) Q is the point representing $\bar{z} = 2 \cos \theta \, e^{-2i\theta}$, being the reflection of P in the x-axis; R is a point representing $z^{1/2}$. One value must lie on OB since its argument will be $\frac{1}{2} . 2\theta = \theta$. (Where does the other value lie? It is on the line RO produced to S where SO = OR.) R lies between O and B if its modulus is less than 1, i.e. if

$$\sqrt{2 \cos \theta} < 1$$

i.e. if $\cos \theta < \frac{1}{2}$
i.e. if $\pi/2 > \theta > \pi/3$.

(iii) $A = e^{3i\alpha}$, $B = e^{i\theta}$; hence
$$z = e^{i\theta} + e^{3i\alpha} = (\cos \theta + \cos 3\alpha) + i(\sin \theta + \sin 3\alpha)$$

i.e. $z - e^{3i\alpha} = e^{i\theta}$ and $|z - e^{3i\alpha}| = |e^{i\theta}| = 1$ so that P lies on a circle centred at A and of radius 1. (Sketch not provided.)

$$OP = |z| = \sqrt{(\cos\theta + \cos 3\alpha)^2 + (\sin\theta + \sin 3\alpha)^2}$$

$$= \sqrt{(\cos^2\theta + \sin^2\theta) + (\cos^2 3\alpha + \sin^2 3\alpha) + 2\cos(3\alpha - \theta)},$$

using the result that $\cos(A - B) \equiv \cos A \cos B + \sin A \sin B$. Now

$$2\cos^2\frac{C}{2} = \cos C + 1$$

and hence

$$OP = \sqrt{2[1 + \cos(3\alpha - \theta)]}$$

$$= \sqrt{4\cos^2\left(\frac{3\alpha - \theta}{2}\right)}$$

$$= 2\cos\left(\frac{3\alpha - \theta}{2}\right).$$

B

Vectors

Example 6

(a) Given the vectors $\mathbf{a} = 3\mathbf{i} + 4\mathbf{j} - \mathbf{k}$, $\mathbf{b} = 2\mathbf{i} - \mathbf{j} + 3\mathbf{k}$, $\mathbf{c} = \mathbf{i} + 5\mathbf{j} - 4\mathbf{k}$, show that they can form the sides of a triangle. Use vector methods to find
   (i) the area of this triangle
   (ii) its angles
   (iii) the lengths of its sides.
(b) Write down the equation of the line through the points with position vectors $\mathbf{f} = (1, -2, -1)$ and $\mathbf{g} = (2, 3, 1)$. Where does this line cut the x-y plane? (EC)

(a) It is sufficient to show that the sum of two of the given vectors is equal to the third. Inspection reveals that $\mathbf{a} = \mathbf{b} + \mathbf{c}$.
   (i) The area of the triangle so formed is given by

$$\Delta = \tfrac{1}{2}|\mathbf{a} \wedge \mathbf{b}| = \tfrac{1}{2}|\mathbf{c} \wedge \mathbf{a}| = \tfrac{1}{2}|\mathbf{b} \wedge \mathbf{c}|$$

(Note that $\mathbf{a} \wedge \mathbf{b} \equiv (\mathbf{b} + \mathbf{c}) \wedge \mathbf{b} = 0 + \mathbf{c} \wedge \mathbf{b}$ etc.)
Now

$$\mathbf{a} \wedge \mathbf{b} = (4 \times 3 - (-1) \times (-1))\mathbf{i} + (-1 \times 2 - 3 \times 3)\mathbf{j}$$
$$+ (3 \times (-1) - 2 \times 4)\mathbf{k}$$

$$= 11\mathbf{i} - 11\mathbf{j} - 11\mathbf{k}$$

Hence

$$\Delta = \frac{1}{2}\sqrt{11^2 + (-11)^2 + (-11)^2} = \frac{11\sqrt{3}}{2}$$

11

(ii) Now $\mathbf{a}.\mathbf{b} = |\mathbf{a}|.|\mathbf{b}|\cos C$, i.e.

$$3 \times 2 + 4 \times (-1) + (-1) \times 3 = \sqrt{26}.\sqrt{14}\cos C$$

so that

$$\cos C = \frac{-1}{\sqrt{26}\sqrt{14}} \quad \text{and} \quad C = 93.00° \qquad \text{(2dp)}$$

Similarly from the relationship $\mathbf{a}.\mathbf{c} = |\mathbf{a}|.|\mathbf{c}|\cos B$ we find that

$$\cos B = \frac{27}{\sqrt{26}\sqrt{42}} \quad \text{and} \quad B = 35.21° \qquad \text{(2dp)}$$

From angle sum it follows that $A = 51.79°$ (2dp)

(It might be argued that since $\cos C$ is so small, it would have been better to have evaluated $\cos A$ and then found C via angle sum. This would have given $\cos A = -0.6186$ which would suggest $A = 128.21°$. However, since the angle C is opposite the longest side it must be the largest angle and hence we choose the supplement of the angle suggested.)

(iii) The lengths of the sides are $|\mathbf{a}| = \sqrt{26} = 5.10$; $|\mathbf{b}| = \sqrt{14} = 3.74$ and $|\mathbf{c}| = \sqrt{42} = 6.48$; all to 2dp.

(b) $\mathbf{g} - \mathbf{f} = (1, 5, 2)$

The equation of the given line is

$$\mathbf{r} = \mathbf{f} + \lambda(\mathbf{g} - \mathbf{f}) = (1, -2, -1) + \lambda(1, 5, 2)$$
$$= (1 + \lambda, -2 + 5\lambda, -1 + 2\lambda)$$

Hence the equation can be written in component form as

$$x = 1 + \lambda, \qquad y = -2 + 5\lambda, \qquad z = -1 + 2\lambda.$$

This line cuts the x–y plane where $z = 0$, i.e.

$$\lambda = \tfrac{1}{2}. \text{ Then } x = \tfrac{3}{2}, \ y = \tfrac{1}{2}.$$

Example 7

(a) Prove that the vectors

$$\mathbf{a} = (3, 1, -2), \qquad \mathbf{b} = (-1, 3, 4), \qquad \mathbf{c} = (4, -2, -6)$$

can form the sides of a triangle, and find the lengths of the medians of this triangle.

(b)   The magnetic flux density **B** can be defined by the equation $\mathbf{F} = q(\mathbf{V} \wedge \mathbf{B})$, where **F** is the force on a charge **q** moving with velocity **V**.

In three experiments it was found that

$$\mathbf{F}/q = -\mathbf{j} - \mathbf{k} \qquad \text{when} \quad \mathbf{V} = \mathbf{i}$$
$$\mathbf{F}/q = \mathbf{i} - 2\mathbf{k} \qquad \text{when} \quad \mathbf{V} = \mathbf{j}$$
$$\mathbf{F}/q = \mathbf{i} + 2\mathbf{j} \qquad \text{when} \quad \mathbf{V} = \mathbf{k}$$

Using these results, calculate **B**.

Find the rate of doing work on the charge when $\mathbf{V} = 2\mathbf{j} + 3\mathbf{k}$.

(CEI)

(a)   Note that $\mathbf{a} = \mathbf{b} + \mathbf{c}$ and hence the three line segments represented by the three vectors can form the sides of a triangle. Refer to Figure 9. Let CE = EA; then one median, BE, is represented by

$$-\mathbf{a} + \tfrac{1}{2}\mathbf{b} = (-3\tfrac{1}{2}, \tfrac{1}{2}, 4)$$

Other medians are

$$\mathbf{a} - \tfrac{1}{2}\mathbf{c} = (1, 2, 1)$$

and

$$\mathbf{c} - \tfrac{1}{2}\mathbf{a} = (2\tfrac{1}{2}, -2\tfrac{1}{2}, -5)$$

These have lengths $\tfrac{1}{2}\sqrt{114}$, $\sqrt{6}$ and $\tfrac{5}{2}\sqrt{6}$ respectively.

(Note that, for example, $\mathbf{c} - \tfrac{1}{2}\mathbf{a} = \tfrac{1}{2}\mathbf{a} - \mathbf{b}$ so that these medians may be obtained by other means; further $\tfrac{1}{2}\mathbf{c} - \mathbf{a}$ can represent the same median as $\mathbf{a} - \tfrac{1}{2}\mathbf{c}$ and the lengths will be calculated as the same.)

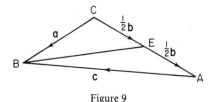

Figure 9

14

(b) Let $\mathbf{B} = (b_1, b_2, b_3)$. Now $\mathbf{i} \wedge \mathbf{B} = -b_3\mathbf{j} + b_2\mathbf{k}$, $\mathbf{j} \wedge \mathbf{B} = b_3\mathbf{i} - b_1\mathbf{k}$
and $\mathbf{k} \wedge \mathbf{B} = -b_2\mathbf{i} + b_1\mathbf{j}$ so that the experimental results become

$$-\mathbf{j} - \mathbf{k} = b_2\mathbf{k} - b_3\mathbf{j}$$
$$\mathbf{i} - 2\mathbf{k} = -b_1\mathbf{k} + b_3\mathbf{i}$$
$$\mathbf{i} + 2\mathbf{j} = b_1\mathbf{j} - b_2\mathbf{i}$$

Hence, comparing coefficients, $b_1 = 2$, $b_2 = -1$, $b_3 = 1$. Therefore

$$\mathbf{B} = (2, -1, 1)$$

Then

$$\mathbf{F} = q(2\mathbf{j} + 3\mathbf{k}) \wedge (2\mathbf{i} - \mathbf{j} + \mathbf{k}) = q(5\mathbf{i} + 6\mathbf{j} - 4\mathbf{k})$$

Rate of doing work is given by

$$\mathbf{F} \cdot \mathbf{V} = q(5\mathbf{i} + 6\mathbf{j} - 4\mathbf{k}) \cdot (2\mathbf{j} + 3\mathbf{k}) = 0 + 12 - 12 = 0.$$

Example 8

(a) Verify the identity

$$(\mathbf{a} \times \mathbf{b}) \cdot (\mathbf{b} \times \mathbf{c}) \times (\mathbf{c} \times \mathbf{a}) = (\mathbf{a} \cdot \mathbf{b} \times \mathbf{c})^2$$

for the vectors

$$\mathbf{a} = \mathbf{i} + \mathbf{j}, \qquad \mathbf{b} = 2\mathbf{j}, \qquad \mathbf{c} = \mathbf{j} + \mathbf{k}$$

and interpret the result geometrically.

(b) If $\mathbf{r}$ is the position vector of a moving point P, find the locus of P when $\mathbf{r}$ satisfies

(i) $\mathbf{r} \cdot \dfrac{d\mathbf{r}}{dt} = 0$ (ii) $\mathbf{r} \times \dfrac{d\mathbf{r}}{dt} = 0$

(c) The sides AB and AC of triangle ABC are divided by points D and E into the ratios shown in Figure 10. Using only vector addition and scaling, find the ratios in which the lines BE and CD divide each other. (*Hint:* Let BG : GE = p: $(1 - p)$, CG : GD = q: $(1 - q)$ and find two equations for p, q from the triangles ABC, GBC, etc.)

(CEI)

The notation $\mathbf{a} \times \mathbf{b}$ is an alternative to $\mathbf{a} \wedge \mathbf{b}$; we use the latter.

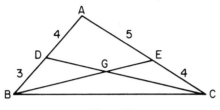

Figure 10

(a) It is easy to verify that

$$\mathbf{a} \wedge \mathbf{b} = 2\mathbf{k}, \qquad \mathbf{b} \wedge \mathbf{c} = 2\mathbf{i}, \qquad \mathbf{c} \wedge \mathbf{a} = -\mathbf{i} + \mathbf{j} - \mathbf{k}$$

Then

$$\mathbf{a} \cdot \mathbf{b} \wedge \mathbf{c} = (\mathbf{i} + \mathbf{j}) \cdot 2\mathbf{i} = 2$$

and this is the volume of the parallelepiped with the origin as one vertex and $\mathbf{a}, \mathbf{b}, \mathbf{c}$ as vectors representing the three edges which meet at O.

$$(\mathbf{a} \wedge \mathbf{b}) \cdot (\mathbf{b} \wedge \mathbf{c}) \wedge (\mathbf{c} \wedge \mathbf{a}) = \begin{vmatrix} 0 & 0 & 2 \\ 2 & 0 & 0 \\ -1 & 1 & -1 \end{vmatrix} = 4$$

and this is the volume of the parallelpiped with the origin as one vertex, $2\mathbf{k}$ and $2\mathbf{i}$ representing two edges meeting there and $-\mathbf{i} + \mathbf{j} - \mathbf{k}$ as the vector representing a third edge. The identity follows since $4 = 2^2$.

(b)   (i) $\mathbf{r} \cdot (\mathrm{d}\mathbf{r}/\mathrm{d}t) = 0 \Rightarrow \mathbf{r} \perp (\mathrm{d}\mathbf{r}/\mathrm{d}t)$ and hence the locus of P is a circle (or, in general, a sphere); refer to Figure 11(a).

     (ii) $\mathbf{r} \wedge (\mathrm{d}\mathbf{r}/\mathrm{d}t) = 0 \Rightarrow \mathbf{r}$ and $\mathrm{d}\mathbf{r}/\mathrm{d}t$ lie in the same straight line, which is the locus of P; see Figure 11(b).

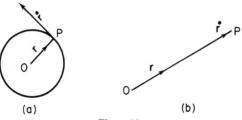

(a)   (b)

Figure 11

16

(c) Let DG = (1 − q)DC and BG = pBE.
In travelling once round the △BAC the net vector displacement is zero. Then

$$\overline{BC} = \overline{BA} + \overline{AC} \qquad (1)$$

Also,

$$\overline{AE} = \tfrac{5}{9}\overline{AC} \quad \text{and} \quad \overline{DA} = \tfrac{4}{7}\overline{BA}.$$

But

$$\overline{BC} = \overline{BG} + \overline{GC} = p\overline{BE} + q\overline{DC} = p(\overline{BA} + \overline{AE}) + q(\overline{DA} + \overline{AC})$$
$$= p(\overline{BA} + \tfrac{5}{9}\overline{AC}) + q(\tfrac{4}{7}\overline{BA} + \overline{AC}) \qquad (2)$$

Comparing coefficients of $\overline{BA}$ and $\overline{AC}$ in (1) and (2), $p + \tfrac{4}{7}q = 1$ and $\tfrac{5}{9}p + q = 1$ so that $p = \tfrac{27}{43}$ and $q = \tfrac{28}{43}$. Hence BG/GE = 27/16 and DG/GC = 15/28.

Example 9

A tetrahedron ABCD is said to be orthocentric if the perpendiculars from vertices A, B, C, and D to the opposite faces intersect in a common point. Figure 12 represents a tetrahedron and shows the perpendiculars from A to BCD and from B to ACD intersecting in the point O.

If ABCD is orthocentric and the position vectors of A, B, C, D relative to O are **a**, **b**, **c**, **d**, respectively, show that

$$\mathbf{a}\,.\,\mathbf{b} = \mathbf{a}\,.\,\mathbf{c} = \mathbf{a}\,.\,\mathbf{d} = \mathbf{b}\,.\,\mathbf{c} = \mathbf{b}\,.\,\mathbf{d} = \mathbf{c}\,.\,\mathbf{d}$$

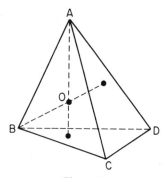

Figure 12

Prove also that opposite edges of this tetrahedron are mutually perpendicular.

Finally, if Q is the point with position vector $\mathbf{q} = \frac{1}{2}(\mathbf{a} + \mathbf{b} + \mathbf{c} + \mathbf{d})$, show that it is equidistant from the vertices A, B, C, and D.　(CEI)

Let the feet of the perpendiculars from A, B, C, D be H, K, L, M, respectively.

If ABCD is orthocentric then AH, BK, CL, DM all intersect in O. Because OA ⊥ BCD, then OA ⊥ to any line in the plane BCD. Therefore OA ⊥ BC and hence $\mathbf{a} \cdot (\mathbf{c} - \mathbf{b}) = 0$ also, OA ⊥ CD so that $\mathbf{a} \cdot (\mathbf{d} - \mathbf{c}) = 0$. Therefore, $\mathbf{a} \cdot \mathbf{b} = \mathbf{a} \cdot \mathbf{c} = \mathbf{a} \cdot \mathbf{d}$.

From the facts that OB ⊥ ACD and OC ⊥ BDA it follows that

$$\mathbf{a} \cdot \mathbf{b} = \mathbf{a} \cdot \mathbf{c} = \mathbf{a} \cdot \mathbf{d} = \mathbf{b} \cdot \mathbf{c} = \mathbf{b} \cdot \mathbf{d} = \mathbf{c} \cdot \mathbf{d} \tag{1}$$

Two opposite edges are AB and CD. Now $\overline{AB} \cdot \overline{CD} = (\mathbf{b} - \mathbf{a}) \cdot (\mathbf{d} - \mathbf{c})$ $= \mathbf{b} \cdot \mathbf{d} - \mathbf{a} \cdot \mathbf{d} - \mathbf{b} \cdot \mathbf{c} + \mathbf{a} \cdot \mathbf{c} = 0$, from (1). Hence

$$AB \perp CD.$$

Similarly,

$$(\mathbf{c} - \mathbf{a}) \cdot (\mathbf{d} - \mathbf{b}) = 0$$

and so AC ⊥ BD; also

$$(\mathbf{d} - \mathbf{a}) \cdot (\mathbf{c} - \mathbf{b}) = 0$$

and therefore AD ⊥ BC. Finally,

$$\begin{aligned}
QA^2 &= (\mathbf{a} - \mathbf{q}) \cdot (\mathbf{a} - \mathbf{q}) = (\tfrac{1}{2}\mathbf{a} - \tfrac{1}{2}\mathbf{b} - \tfrac{1}{2}\mathbf{c} - \tfrac{1}{2}\mathbf{d})^2 \\
&= \tfrac{1}{4}(\mathbf{a} - \mathbf{b} - \mathbf{c} - \mathbf{d})^2 = \tfrac{1}{4}(\mathbf{a}^2 + \mathbf{b}^2 + \mathbf{c}^2 + \mathbf{d}^2) \\
&\quad - \tfrac{1}{2}(\mathbf{a} \cdot \mathbf{b} + \mathbf{a} \cdot \mathbf{c} + \mathbf{a} \cdot \mathbf{d}) \\
&\quad + \tfrac{1}{2}(\mathbf{a} \cdot \mathbf{c} + \mathbf{b} \cdot \mathbf{d} + \mathbf{c} \cdot \mathbf{d}) \\
&= \tfrac{1}{4}(\mathbf{a}^2 + \mathbf{b}^2 + \mathbf{c}^2 + \mathbf{d}^2), \text{ using (1)}
\end{aligned}$$

Arguments of symmetry will show that

$$QA^2 = QB^2 = QC^2 = QD^2 = \tfrac{1}{4}(a^2 + b^2 + c^2 + d^2)$$

so that Q (the circumcentre of ABCD) is equidistant from the four vertices.

18

Example 10

(a) Verify for the vectors

$$\mathbf{a} = \left(\frac{3}{5}, 0, \frac{4}{5}\right), \mathbf{b} = \left(\frac{4}{5}, \frac{3}{5}, 0\right) \quad \text{and} \quad \mathbf{c} = \left(0, \frac{4}{5}, -\frac{3}{5}\right)$$

the following identities:

(i) $\mathbf{a} \times (\mathbf{b} \times \mathbf{c}) = (\mathbf{a} \cdot \mathbf{c})\mathbf{b} - (\mathbf{a} \cdot \mathbf{b})\mathbf{c}$

(ii) $(\mathbf{a} \times \mathbf{b}) \cdot (\mathbf{a} \times \mathbf{c}) = \mathbf{b} \cdot \mathbf{c} - (\mathbf{a} \cdot \mathbf{b})(\mathbf{a} \cdot \mathbf{c})$

If $\overline{OA} = \mathbf{a}$, $\overline{OB} = \mathbf{b}$, $\overline{OC} = \mathbf{c}$ and $<BOC = \alpha$, $<COA = \beta$, $<AOB = \gamma$, determine α, β, γ and the areas of the triangles BOC, COA, and AOB.

(b) If $\mathbf{a}$ and $\mathbf{b}$ are non-parallel vectors, explain why it is possible to choose scalar constants p, q, r such that any vector $\mathbf{x}$ can be expressed in the form

$$\mathbf{x} = p\mathbf{a} + q\mathbf{b} + r(\mathbf{a} \times \mathbf{b})$$

Hence, or otherwise, solve for $\mathbf{x}$ the vector equation

$$\mathbf{x} + (\mathbf{x} \times \mathbf{a}) = \mathbf{b} \qquad \text{(CEI)}$$

(a) It is straightforward to show that

$$\mathbf{a} \wedge \mathbf{b} = \left(-\frac{12}{25}, \frac{16}{25}, \frac{9}{25}\right); \qquad \mathbf{b} \wedge \mathbf{c} = \left(-\frac{9}{25}, \frac{12}{25}, \frac{16}{25}\right);$$

$$\mathbf{a} \wedge \mathbf{c} = \left(-\frac{16}{25}, \frac{9}{25}, \frac{12}{25}\right); \qquad \mathbf{a} \wedge (\mathbf{b} \wedge \mathbf{c}) = \left(-\frac{48}{125}, -\frac{84}{125}, \frac{36}{125}\right);$$

$$(\mathbf{a} \wedge \mathbf{b}) \cdot (\mathbf{a} \wedge \mathbf{c}) = 444/625$$

$$\mathbf{a} \cdot \mathbf{c} = -\frac{12}{25}; \qquad \mathbf{a} \cdot \mathbf{b} = \frac{12}{25}; \qquad \mathbf{b} \cdot \mathbf{c} = \frac{12}{25}.$$

(i) RHS $= -\frac{12}{25}\left(\frac{4}{5}, \frac{3}{5}, 0\right) - \frac{12}{25}\left(0, \frac{4}{5}, -\frac{3}{5}\right) = \left(-\frac{48}{125}, -\frac{84}{125}, \frac{36}{125}\right) =$ LHS

(ii) RHS $= \frac{12}{25} + \frac{12}{25} \cdot \frac{12}{25} = \frac{300}{625} + \frac{144}{625} =$ LHS

Since $|\mathbf{a}| = |\mathbf{b}| = |\mathbf{c}| = 1$ then

$$|\mathbf{b} \cdot \mathbf{c}| = |\mathbf{b}| \cdot |\mathbf{c}| \cdot \cos\alpha \Rightarrow \frac{12}{25} = 1.1. \cos\alpha \Rightarrow \underline{\alpha = 61°19'}$$

Similarly,

$$\beta = 118°41', \gamma = 61°19'$$

Now,

$$|\mathbf{b} \wedge \mathbf{c}| = \frac{\sqrt{481}}{25} = 2\Delta BOC \quad \text{and}$$

$$2\Delta AOC = |\mathbf{c} \wedge \mathbf{a}| = \frac{\sqrt{481}}{25} = |\mathbf{a} \wedge \mathbf{b}| = 2\Delta AOB$$

Hence

$$\Delta BOC = \Delta COA = \Delta AOB = 0.439 \ (3dp)$$

(b) Since $(\mathbf{a} \wedge \mathbf{b})$ is perpendicular to both $\mathbf{a}$, $\mathbf{b}$ then these three vectors form a linearly independent set. Then any vector $\mathbf{x}$ can be expressed as $\mathbf{x} = p\mathbf{a} + q\mathbf{b} + r(\mathbf{a} \wedge \mathbf{b})$ where p, q, r are constants. The equation

$$\mathbf{x} + \mathbf{x} \wedge \mathbf{a} = \mathbf{b} \Rightarrow p\mathbf{a} + q\mathbf{b} + r(\mathbf{a} \wedge \mathbf{b}) + q(\mathbf{b} \wedge \mathbf{a}) + r(\mathbf{a} \wedge \mathbf{b}) \wedge \mathbf{a} = \mathbf{b} \quad (1)$$

(Note that $\mathbf{a} \wedge \mathbf{a} = 0$). But $(\mathbf{a} \wedge \mathbf{b}) \wedge \mathbf{a} = -\mathbf{a} \wedge (\mathbf{a} \wedge \mathbf{b}) = -(\mathbf{a}.\mathbf{b}) + (\mathbf{a}.\mathbf{a})\mathbf{b}$ by (i) and $\mathbf{b} \wedge \mathbf{a} = -\mathbf{a} \wedge \mathbf{b}$.

Comparing coefficients in (1)

$$\mathbf{a}: \ p - r(\mathbf{a}.\mathbf{b}) = 0$$
$$\mathbf{b}: \ q + ra^2 \ \ \ \ = 1$$
$$\mathbf{a} \wedge \mathbf{b}: \ r - q \ \ \ \ = 0$$

Therefore,

$$\mathbf{x} = \frac{1}{(1 + a^2)} \{(\mathbf{a}.\mathbf{b})\mathbf{a} + \mathbf{b} + (\mathbf{a} \wedge \mathbf{b})\}$$

C

Linear Algebra

Example 11

(a) If B is a 2×2 matrix with no zero elements, verify that

$$|BB^{\mathrm{T}}| = |B|^2$$

(b) If A and C are two matrices such that $AC = A$ and $CA = C$, show that $A^2 = A$ and $C^2 = C$.

(c) In the following circuit relations

$$\mathbf{i} = C\mathbf{u}, \qquad \mathbf{u} = Z^{-1}\mathbf{v}, \qquad \mathbf{v} = C^{\mathrm{T}}\mathbf{e}, \qquad Z = C^{\mathrm{T}}WC$$

$\mathbf{i}$, $\mathbf{u}$, $\mathbf{v}$, and $\mathbf{e}$ are column vectors and C, Z, W are non-singular matrices.

Given that for any non-singular matrices P, Q and R,

$$(PQR)^{-1} = R^{-1}Q^{-1}P^{-1} \quad \text{and} \quad (P^{\mathrm{T}})^{-1} = (P^{-1})^{\mathrm{T}},$$

show that the four circuit relations reduce to

$$\mathbf{i} = W^{-1}\mathbf{e}. \tag{CEI}$$

(a) Let $B = \begin{bmatrix} a & b \\ c & d \end{bmatrix}$ then $BB^{\mathrm{T}} = \begin{bmatrix} a^2 + b^2 & ac + bd \\ ac + bd & c^2 + d^2 \end{bmatrix}$

Hence

$$|BB^T| = (a^2 + b^2)(c^2 + d^2) - (ac + bd)(ac + bd)$$
$$= a^2c^2 + b^2c^2 + a^2d^2 + a^2d^2 + b^2d^2$$
$$- a^2c^2 - abcd - abcd - b^2d^2$$
$$= a^2d^2 + b^2c^2 - 2abcd$$
$$= (ad - bc)^2 = |B|^2, \text{ as required.}$$

(b)
$$A^2 = (AC)A = A(CA) = AC = A$$
$$C^2 = (CA)C = C(AC) = CA = C$$

(c)
$$\mathbf{i} = CZ^{-1}C^T\mathbf{e}$$
$$= C[C^{-1}W^{-1}(C^T)^{-1}]C^T\mathbf{e}$$
$$= (CC^{-1})W^{-1}(C^T)^{-1}C^T\mathbf{e}$$
$$= W^{-1}\mathbf{e}$$

Note that the result $(P^T)^{-1} = (P^{-1})^T$ was not used.

Example 12

Use elimination methods to obtain expressions for x, y, and z in terms of the parameters α and β from the linear equations

$$x + y + z = 8$$
$$\alpha x + 2y + \beta z = 4$$
$$\alpha x + \beta y + \alpha z = 4$$

For what ranges of values of α and β do the following conditions occur?

(i)   The equations have a unique solution.
(ii) The equations have a non-unique solution.
(iii) The equations are inconsistent.
(iv) The equations are consistent.

Find the solution when $\alpha = 0.6$ and $\beta = 0.5$.   (CEI)

The augmented matrix for the system is

$$\begin{bmatrix} 1 & 1 & 1 & \vdots & 8 \\ \alpha & 2 & \beta & \vdots & 4 \\ \alpha & \beta & \alpha & \vdots & 4 \end{bmatrix}$$

Subtracting $\alpha \times$ row 1 from rows 2 and 3 we obtain the augmented matrix

$$\begin{bmatrix} 1 & 1 & 1 & \vdots & 8 \\ 0 & 2-\alpha & \beta-\alpha & \vdots & 4-8\alpha \\ 0 & \beta-\alpha & 0 & \vdots & 4-8\alpha \end{bmatrix}$$

From the third row we find that

$$y = \left(\frac{4-8\alpha}{\beta-\alpha}\right), \quad \alpha \neq \beta;$$

substitution into the second row produces

$$z = \frac{(4-8\alpha)(\beta-2)}{(\beta-\alpha)^2}, \quad \alpha \neq \beta;$$

and from the first row

$$x = 8 - \frac{(4-8\alpha)}{(\beta-\alpha)}\left(1+\left[\frac{\beta-2}{\beta-\alpha}\right]\right), \quad \alpha \neq \beta$$

$$= \frac{8\beta(\beta-1)+8-12\alpha}{(\beta-\alpha)^2}, \quad \alpha \neq \beta$$

(i) There is a unique solution if $\alpha \neq \beta$

(Then rank $A = 3$ or $|A| = -(\beta-\alpha)^2 \neq 0$)

(ii) If $\beta = \alpha$ there is no unique solution and the last row is inconsistent if $\alpha \neq \frac{1}{2}$; therefore the equations are inconsistent if $\alpha = \beta \neq \frac{1}{2}$.

(iii) Whereas if $\alpha = \beta = \frac{1}{2}$ there is a non-unique solution given by $\frac{3}{2}y = 0$, $x+y+z = 8$, i.e.

$$y = 0 \quad \text{and} \quad x = 8-z$$

(iv) Finally, the equations are consistent if

$$\text{either } \alpha \neq \beta$$
$$\text{or } \alpha = \beta = \frac{1}{2}.$$

When $\alpha = 0.6$, $\beta = 0.5$, the solution is found from the general formulae to be $x = -120$, $y = 8$, $z = 120$.

Example 13

(a) Show that the equations

$$5x_1 + x_2 + 7x_3 = 0$$
$$4x_1 + x_2 + 11x_3 = 0$$
$$x_1 - x_2 + 5x_3 = 0$$

are linearly dependent, and that the solution set can be expressed in the form $(-2x_3, 3x_3, x_3)$, with x_3 arbitrary.

(b) Consider the system of equations

$$\begin{bmatrix} 5 & 1 & 7 \\ 4 & -1 & 11 \\ 1 & -1 & a \end{bmatrix} \begin{bmatrix} x_1 \\ x_2 \\ x_3 \end{bmatrix} = \begin{bmatrix} 0 \\ 2 \\ 1 \end{bmatrix}$$

   (i) Show that the equations are inconsistent when $a = 5$.
   (ii) Using Gauss elimination in the case when $a = 5.01$, obtain the result $x_3 \simeq -33$. (CEI)

With a view to later parts of the question we consider the more general system of equations where the right-hand sides are replaced by α, β, and γ respectively.

Interchanging the first and third equations leads to the general augmented matrix

$$\begin{bmatrix} 1 & -1 & 5 & \vdots & \gamma \\ 4 & -1 & 11 & \vdots & \beta \\ 5 & 1 & 7 & \vdots & \alpha \end{bmatrix}$$

This can be reduced to the following by Gauss operations
Row $2 - 4 \times$ Row 1 and Row $3 - 5 \times$ Row 1

$$\begin{bmatrix} 1 & -1 & 5 & \vdots & \gamma \\ 0 & 3 & -9 & \vdots & \beta - 4\gamma \\ 0 & 6 & -18 & \vdots & \alpha - 5\gamma \end{bmatrix}$$

(a) If $\alpha = \beta = \gamma = 0$ then row $3 = 2 \times$ row 2 and the equations are therefore linearly dependent. The reduced second equation gives

$$3x_3 = x_2$$

and the first equation in this new order then produces

$$x_1 = x_2 - 5x_3 = -2x_3.$$

Hence $(-2x_3, 3x_3, x_3)$ is the general solution.

(b) If $\alpha = 0$, $\beta = 2$, $\gamma = 1$ and

  (i) $a = 5$ then the equations reduce to

$$\begin{bmatrix} 1 & -1 & 5 & \vdots & 1 \\ 0 & 3 & -9 & \vdots & -2 \\ 0 & 6 & -18 & \vdots & -5 \end{bmatrix}$$

and by comparing the last two rows it is seen that the original equations are inconsistent.

  (ii) Gauss elimination leads to

$$\begin{bmatrix} 1 & -1 & 5.01 & \vdots & 1 \\ 0 & 3 & -9.04 & \vdots & -2 \\ 0 & 6 & -18.05 & \vdots & -5 \end{bmatrix}$$

Whereupon subtracting twice row 2 from row 3 leads to

$$\begin{bmatrix} 1 & -1 & 5.01 & \vdots & 1 \\ 0 & 3 & -9.04 & \vdots & -2 \\ 0 & 0 & -0.03 & \vdots & -1 \end{bmatrix}$$

i.e. $x_3 \simeq -33$.

(In fact, the system is ill-conditioned since $|A| \simeq -0.09$, which is very small compared with the elements of A.)

Example 14

(a) State any four properties of determinants. Hence, or otherwise, factorize the determinant

$$\begin{vmatrix} a+1 & b+1 & c+1 \\ a & b & c \\ a(a+1) & b(b+1) & c(c+1) \end{vmatrix}$$

and deduce the value of

$$\begin{vmatrix} 1.2 & 1.3 & 1.4 \\ 0.2 & 0.3 & 0.4 \\ 0.24 & 0.39 & 0.56 \end{vmatrix}$$

(b) Given that $bc = 1$, what values of a will make the matrix

$$A(a, b, c) = \begin{bmatrix} a & b & 0 & 0 \\ c & a & b & 0 \\ 0 & c & a & b \\ 0 & 0 & c & a \end{bmatrix} \text{ singular?}$$

Reduce the matrix $A(1.6, 1.0, 1.0)$ to triangular form and comment briefly on the result. (CEI)

(a) The first part is bookwork. For example;

   (i) If two rows (or columns) of a determinant are equal, its value is zero.

  (ii) If a row (or column) is multiplied by a constant λ, then the value of the determinant is multiplied by λ.

 (iii) If two rows (or columns) of a determinant are interchanged then the value of the determinant is multiplied by (-1).

 (iv) If a row (or column) is increased by adding a multiple of another row (or column), the value of the determinant is not changed.

Now
$$\begin{vmatrix} a+1 & b+1 & c+1 \\ a & b & c \\ a(a+1) & b(b+1) & c(c+1) \end{vmatrix}$$

$$= \begin{vmatrix} a & b & c \\ a & b & c \\ a(a+1) & b(b+1) & c(c+1) \end{vmatrix}$$

$$+ \begin{vmatrix} 1 & 1 & 1 \\ a & b & c \\ a(a+1) & b(b+1) & c(c+1) \end{vmatrix} \text{ by (iv) on rows}$$

and

$$\begin{vmatrix} a & b & c \\ a & b & c \\ a(a+1) & b(b+1) & c(c+1) \end{vmatrix} = \begin{vmatrix} a & b & c \\ a & b & c \\ a^2 & b^2 & c^2 \end{vmatrix}$$

$$+ \begin{vmatrix} a & b & c \\ a & b & c \\ a & b & c \end{vmatrix}$$

therefore

$$D(a,b,c) = \begin{vmatrix} a+1 & b+1 & c+1 \\ a & b & c \\ a(a+1) & b(b+1) & c(c+1) \end{vmatrix}$$

$$= \begin{vmatrix} a & b & c \\ a & b & c \\ a^2 & b^2 & c^2 \end{vmatrix} + \begin{vmatrix} a & b & c \\ a & b & c \\ a & b & c \end{vmatrix}$$

$$+ \begin{vmatrix} 1 & 1 & 1 \\ a & b & c \\ a^2 & b^2 & c^2 \end{vmatrix} + \begin{vmatrix} 1 & 1 & 1 \\ a & b & c \\ a & b & c \end{vmatrix}$$

The first, second, and fourth determinants are each zero since they have two identical rows. Therefore

$$D(a,b,c) = \begin{vmatrix} 1 & 1 & 1 \\ a & b & c \\ a^2 & b^2 & c^2 \end{vmatrix}$$

$$= (bc^2 - b^2c) - (ac^2 - a^2c) + (ab^2 - a^2b) \qquad (1)$$

It is not too easy to produce a simple factorization from here. We note that if $a = b$, two columns of D are equal and hence $D = 0$; therefore $(a - b)$ is a factor of D. In the same way it can be shown that $(b - c)$ and $(c - a)$ are also factors so that $D = K(a - b)(b - c)(c - a)$; by comparing the coefficient of say a^2c with that in (1) it is seen that $K = 1$. Hence

$$D(a,b,c) = \underline{(a - b)(b - c)(c - a)}.$$

However, this approach applied directly to the original form of D would give the result directly. Note that

$$(a - b)(b - c)(c - a) \equiv abc - b^2c - a^2b + ab^2$$
$$- ac^2 + bc^2 + a^2c - abc,$$

which reduces to (i).

Finally, we evaluate the given determinant as

$$D(0.2, 0.3, 0.4) = (-0.1)(-0.1)(0.2) = \underline{0.002}.$$

(b) $|A| = a \begin{vmatrix} a & b & 0 \\ c & a & b \\ 0 & c & a \end{vmatrix} - b \begin{vmatrix} c & b & 0 \\ 0 & a & b \\ 0 & c & a \end{vmatrix},$

expanding along top row,

$$= a\left\{ a\begin{vmatrix} a & b \\ c & a \end{vmatrix} - b\begin{vmatrix} c & b \\ 0 & a \end{vmatrix}\right\} - bc\begin{vmatrix} a & b \\ c & a \end{vmatrix}$$

$$= (a^2 - bc)(a^2 - bc) - ab \cdot ac$$

$$= a^4 - 3a^2 + 1, \qquad \text{since } bc = 1.$$

Hence, A is singular if $a^2 = \frac{1}{2}(3 \pm \sqrt{5})$

$$\text{or } \underline{a = \pm 1.618, \pm 0.618} \text{ (3dp)}$$

By Gauss elimination, we can reduce

$$\begin{bmatrix} 1.6 & 1 & 0 & 0 \\ 1 & 1.6 & 1 & 0 \\ 0 & 1 & 1.6 & 1 \\ 0 & 0 & 1 & 1.6 \end{bmatrix} \text{ to } \begin{bmatrix} 1.6 & 1 & 0 & 0 \\ 0 & 0.975 & 1 & 0 \\ 0 & 0 & 0.524 & 1 \\ 0 & 0 & 0 & -0.142 \end{bmatrix}$$

Then, $|A| = (1.6)(0.975)(0.524)(-0.142) \simeq -0.116.$ A is, therefore, nearly singular (and would have been if $a = 1.618$).

D

Curve Sketching

If a *sketch* is asked for then plotting should be used as a last resort. The main features of the curve can be deduced from its equation almost by inspection, with a few extra details obtained by differentiating to find the gradient of the curve.

Some of the features to look for are listed below.

1. *Symmetry*. The curve is symmetrical about the x-axis if its equation does not change when y is replaced by $-y$; for example, y occurs only as even powers. The curve is symmetrical about the y-axis if replacing x in the equation by $-x$ produces no change. There is symmetry about $y = x$ if interchanging x and y leaves the equation unaltered. If replacing x by $-x$ changes the sign of y, then the curve is skew-symmetrical about the x-axis; for example $y = x^3$.

2. *Forbidden regions*. Check to see whether some values of x or y are not possible. For example the equation $y^2 = x - 1$ implies that $x \geqslant 1$ since y^2 cannot be negative.

3. *Intercepts*. Put $y = 0$ to find the curve crosses the x-axis and put $x = 0$ to find where it crosses the y-axis. Note that a double root may imply that the axis is a tangent, e.g. $y = x^2$.

4. *Behaviour for large $|x|$*. When $|x|$ becomes large (assuming the equation of the curve indicates that this is possible) it may be possible to find a simpler equation which approximates the behaviour of the curve. For example, the equation

$$y^3 = -x^3 + 9x^2$$

can be written as

$$y^3 = -x^3\left(1 - \frac{9}{x}\right)$$

so that

$$y = -x\left(1 - \frac{9}{x}\right)^{1/3}$$

$$\simeq -x + 3$$

for large $|x|$, using the binomial theorem. The line $y = -x + 3$ is an *asymptote* of the curve.

5. *Behaviour near the origin.* If the curve passes through the origin, its behaviour nearby may be found by a simplifying approximation. For example the equation $y^2(1 - x) = x^2(1 + x)$ can be approximated by $y^2 = x^2$ when $|x|$ is small.

   In the form $(y - x)(y + x) = 0$ it can be seen to be two straight lines through the origin.

6. *Stationary points.* Differentiating the equation to find dy/dx allows the determination of local maximum, minimum, points of inflection etc. If the formula for dy/dx is a fraction, then where the numerator is zero we have a horizontal gradient and where the denominator is zero we have a vertical gradient. Where both are zero simultaneously, further investigation is needed.

7. *Asymptotes.* A curve can have asymptotes other than when $|x|$ is large. Refer to the examples which follow for illustrations of this.

Example 15

Given that

$$\frac{x}{\sqrt{1 - x^2/c^2}}, \qquad c \text{ constant,}$$

sketch the following curves and indicate their important features:

(i) $y(x)$ (ii) $y'(x)$ (iii) $\dfrac{1}{c}\displaystyle\int_0^x y(t)\, dt = \mu(x).$

Deduce the mean value of $y(x)$ on the interval $[0, c]$. How do the functions above behave as $c \to \infty$? (CEI)

(i) Since $y(-x) = -y(x)$ there is *skew-symmetry* about the y-axis.
Intercepts with axes occur only at origin.
Range. Curve exists only when $-c \leqslant x \leqslant c$, i.e. $|x| \leqslant c$
Asymptotes. As $x \to \pm c$, $y \to \pm \infty$, hence the lines $x = \pm c$ are

asymptotes. Since

$$y = x\left[1 + \frac{x^2}{2c^2} + 0(x^4)\right]$$

for $|x| < c$, then near the origin, $y/x \simeq 1+$ so that $y \simeq x$. See Figure 13(a) where the asymptotes are shown as dotted lines.

(ii) $y'(x) = 1/(1 - x^2/c^2)^{3/2} = y^3/x^3$.

Symmetry about y-axis, since $y'(-x) = y'(x)$
Intercepts with axes do not occur.
Range. Curve exists only when $|x| < c$.
Asymptotes. As $x \to \pm c$, $y' \to \infty$. Since

$$y' = 1 + \frac{3x^2}{2c^2} + 0(x^4)$$

for $|x| < c$, then near $x = 0$, $y' \simeq 1+$. At $x = 0$, y' has its lowest value by inspection of the equation for $y'(x)$; there is no need to find $y''(x)$ explicitly. Refer to Figure 13(b).

(iii) $\mu(x) = c[1 - \sqrt{1 - x^2/c^2}] = c(1 - x/y)$

Symmetry $\mu(-x) = \mu(x)$ so there is symmetry about y-axis.
Intercept with axes only at origin.
Range. Curve exists only for $|x| < c$. Note that $\mu(\pm c) = c$.
Asymptotes. None.

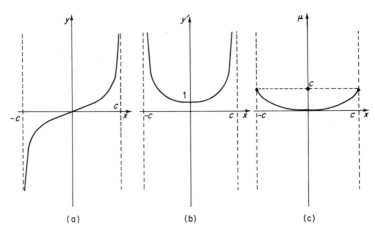

(a)   (b)   (c)

Figure 13

Since

$$\mu = \frac{x^2}{2c} + \frac{x^4}{8c^3} + 0(x^6) \qquad (*)$$

for $|x| < c$, then near $x = 0$, $\mu \simeq x^2/2c +$. See Figure 13(c). The sketches can be given more detail by calculating values of each function at $x = \frac{1}{2}c$ etc. Mean value of $y(x)$ on $[0, c] = \mu(c) = c$. As $c \to \infty$, $y \to x$, $y' \to 1$, $\mu \to x^2/2c$.

The first two results can be obtained directly from the equations of the respective functions and the third follows from (*).

Example 16

(a) The decay of a vortex in a viscous fluid is given by

$$U = \frac{K}{2\pi r}\left\{1 - \exp\left(-\frac{r^2}{4\nu t}\right)\right\}$$

where U is the tangential velocity at a point distant r from the origin at time t, and K, ν are constants.

Sketch the curve of U against r when $\nu t = 1$, describing the main features.

(b) Show that $4xy = x^2 + y^2$ represents a pair of straight lines which are symmetrically placed about the lines $y = \pm x$.

Show further that the curve $4 \sin xy = x^2 + y^2$ has similar symmetry but is closed and bounded by a circle of radius 2. Hence sketch this curve. (CEI)

(a) With the condition $\nu t = 1$, the curve has a simplified equation, namely

$$U = \frac{K}{2\pi r}\left[1 - \exp(-r^2/4)\right]$$

(i) Since $U(-r) = -U(r)$ there is skew-symmetry about the r-axis. Since $r < 0$ is physically meaningless we concentrate on $r \geqslant 0$.

(ii) Since $0 < \exp(-r^2/4) < 1$ then $0 < 2\pi rU/K < 1$, i.e.

$$U > 0 \quad \text{for} \quad r > 0.$$

(iii) As $r \to \infty$ $U \to (K/2\pi r) \to 0+$; hence the r-axis is an asymptote.

(iv) For small r,

$$U = \frac{K}{2\pi r}\left[\frac{r^2}{4} + 0(r^4)\right] \simeq \frac{Kr}{8\pi} \to 0+$$

as $r \to 0+$. It makes sense to define $U = 0$ when $r = 0$.

(v) $U'(r) = \dfrac{K}{2\pi}\left[\dfrac{1}{2}\exp(-r^2/4) - \dfrac{1}{r^2}(1 - \exp(-r^2/4))\right]$

Then $U' \to 0$ as $r \to \infty$. Since $1 - \exp(-r^2/4) = 1 - 1 + (r^2/4) + 0(r^4)$, then as $r \to 0+$,

$$U' \to \frac{K}{2\pi}\left(\frac{1}{2} - \frac{1}{4}\right) = \frac{K}{8\pi}.$$

Now

$$U' = 0 \Rightarrow (1 + r^2/2) = \exp(r^2/4) \simeq 1 + \frac{r^2}{4} + \frac{r^4}{32}$$

There is a root of this equation at $r \simeq \sqrt{8}$ which is clearly a local maximum. See Figure 14.

(b) The symmetry about $y = x$ may be demonstrated in a number of ways. The easiest is to show that if the point (a, b) lies on the curve then so does the point (b, a); that is, the equation of the curve is not affected by interchanging x and y. This is clearly true for both curves.

In Figure 15 let P have coordinates $(a, b) = r(\cos \theta + i \sin \theta)$ where $\theta = (\pi/4) + \alpha$. Consider the point Q which is the mirror im-

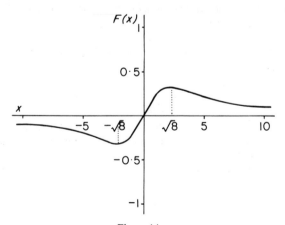

Figure 14

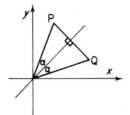

Figure 15

age of P in the line $y = x$. In the polar notation its modulus is also r and its argument is $(\pi/4 - \alpha)$. Then

$$Q = r(\cos(\pi/4 - \alpha) + i \sin(\pi/4 - \alpha))$$
$$= r(\sin(\pi/4 + \alpha) + i \cos(\pi/4 + \alpha))$$
$$= r(\sin \theta + i \cos \theta) = (b, a).$$

An alternative method of demonstrating symmetry is based on the polar substitution $x = r \cos \theta$, $y = r \sin \theta$; the equation of the first curve reduces to

$$\sin 2\theta = \frac{1}{2} \quad \text{so that} \quad \theta = \frac{\pi}{12} \quad \text{or} \quad \frac{5\pi}{12}$$

which is clearly symmetrical about $\theta = \pi/4$. However, to obtain the equations of the two straight lines suggested in the question it is probably easier to substitute $y = kx$ (since the origin lies on the curves) and find that $k = 2 \pm \sqrt{3}$ so that the lines are

$$y = (2 + \sqrt{3})x \quad \text{and} \quad y = (2 - \sqrt{3})x.$$

Symmetry about $y = -x$ can be demonstrated by showing that if (a, b) lies on the curve then so does $(-b, -a)$.

It is again clearly seen that both curves have this property.

In the case of the second curve, $\sin xy = \frac{1}{4} r^2 \geqslant 0$ so that x and y must have the same sign; the curve can only exist in the first and third quadrants.

Also since $-1 \leqslant \sin \leqslant 1$ then the curve is bounded by the circle $x^2 + y^2 = 4$. The line $y = x$ is intersected where

$$4 \sin x^2 = 2x^2 \quad \text{or} \quad x^2 \simeq 1.896$$

i.e. $x \simeq \pm 1.377$

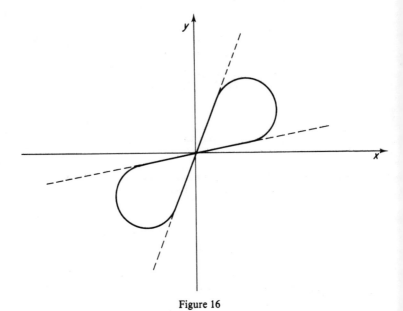

Figure 16

Note that near the origin $\sin xy \simeq xy$ and the present curve is approximated by the straight lines of the first curve, which is shown by broken lines in Figure 16.

Example 17

Sketch the curve $y^2 = 4x^2(1 - x^2)$, indicating characteristics such as symmetry, behaviour at the origin and turning points.

By substituting $x = \sin t$ in the given equation, show that the curve may be represented parametrically by

$$x = \sin t, \quad y = \sin 2t.$$

Indicate on your sketch the points generated by the range of values from $t = 0$ to $t = 2\pi$ at intervals of $\pi/4$.   (CEI)

There is symmetry about both axes, and intercepts with axes at origin, at $(1, 0)$ and at $(-1, 0)$. Near the origin, $y^2 \simeq 4x^2$ i.e. $y = \pm 2x$. We require $x^2 \leqslant 1$, hence curve defined only for $-1 \leqslant x \leqslant 1$. If $x = \sin t$, it follows that $y = \sin 2t$ and hence $-1 \leqslant y \leqslant 1$.

There are no asymptotes. It can be shown that

$$y' = \pm \frac{2(1 - 2x^2)}{\sqrt{1 - x^2}}$$

so $y' = 0$ where $x^2 = \pm (1/\sqrt{2})$. Investigating the sign of y' on either side of these turning points and remembering the double symmetry, we conclude that there are local maxima at $\left(\pm \dfrac{1}{\sqrt{2}}, 1 \right)$ and local minima at $\left(\pm \dfrac{1}{\sqrt{2}}, -1 \right)$.

As $x \to \pm 1$, $y' \to \pm \infty$, indicating vertical gradients.

Table 1 helps us to construct Figure 17. Values of t are marked on the sketch.

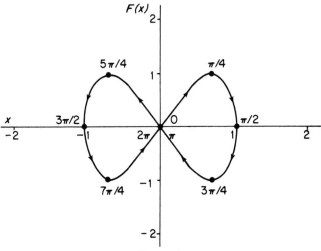

Figure 17

Table 1

t	0	$\dfrac{\pi}{4}$	$\dfrac{\pi}{2}$	$\dfrac{3\pi}{4}$	π	$\dfrac{5\pi}{4}$	$\dfrac{3\pi}{2}$	$\dfrac{7\pi}{4}$	2π
x	0	$\dfrac{1}{\sqrt{2}}$	1	$\dfrac{1}{\sqrt{2}}$	0	$\dfrac{-1}{\sqrt{2}}$	-1	$\dfrac{-1}{\sqrt{2}}$	0
y	0	1	0	-1	0	1	0	-1	0

Example 18

Sketch the following curves and in each case show turning points, points of inflexion, asymptotes, and any other features of interest.

(i) $x^2 y^2 = 4$ and $y^2 = x^2 + (4/x^2)$ (on the same diagram).

(ii) $y = \dfrac{1}{\sigma\sqrt{(2\pi)}} \exp[-(x-\mu)^2/2\sigma^2]$

where μ and σ are constants.   (CEI)

(Outline solution only)

(i) (a) $x^2 y^2 = 4$.
     Symmetry about both axes.
     No intercepts with axes.
     Axes are asymptotes.
     No local max/min.
     $y' = \pm 2/x^2$. The curve is shown with solid lines in Figure 18.

     (b) $y^2 = x^2 + 4/x^2$.
     Symmetry about both axes.
     No intercepts.

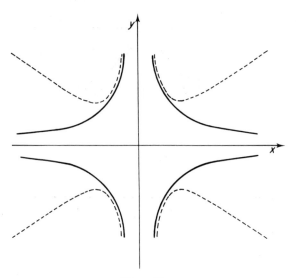

Figure 18

Require $y^2 \geqslant 4$,
but min $y^2 = 4$ where $x = \pm\sqrt{2}$
Now y-axis is an asymptote.
Since $x \to \infty \Rightarrow y^2 \simeq x^2$ then $y = \pm x$ is also an asymptote.

$$y' = \pm \frac{(x^4 - 4)}{x^2\sqrt{x^4 + 4}}, \ y'' = \frac{-64}{x^3(x^4 + 4)^{3/2}}$$

T.P. at $x = \pm\sqrt{2}$, $y = \pm 2$ (local max, min respectively). For small $|x|$, $y^2 \simeq 4/x^2$. (This curve is shown with dashes.)

(ii) Symmetry about $x = \mu$. Intercepts y-axis only at

$$y = \frac{1}{\sigma\sqrt{2\pi}} \exp(-\mu^2/2\sigma^2)$$

As $|x| \to \infty$, $y \to 0+$ hence the x-axis is an asymptote.

$$y' = \frac{1}{\sigma\sqrt{2\pi}} \exp[-(x-\mu)^2/2\sigma^2] \left[\frac{-(x-\mu)}{\sigma^2} \right]$$

This yields a local maximum at $x = \mu$, where

$$y = \frac{1}{\sigma\sqrt{2\pi}}$$

Also,

$$y'' = \frac{1}{\sigma\sqrt{2\pi}} \exp[-(x-\mu)^2/2\sigma^2] \left[\frac{(x-\mu)^2}{\sigma^4} - \frac{1}{\sigma^2} \right]$$

hence there are points of inflexion at $x = \mu \pm \sigma$. Figure 19 depicts the required sketch.

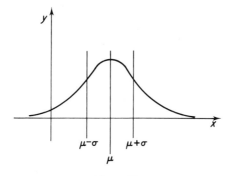

Figure 19

Example   19

Sketch the following curves, pointing out features of interest, including asymptotes and turning points.

(i) $y = x\sqrt{\left(\dfrac{x}{x-1}\right)}$.

(ii) $x^3 + y^3 = 3xy$ (*Hint*: put $y = tx$ and work in terms of t.)

(CEI)

(i)   No symmetry. Intercept with axes only at origin. y is not real if $0 < x < 1$ since $x/(x-1)$ is negative for this range of values of x. Since

$$y = x\left(1 - \frac{1}{x}\right)^{-1/2},$$

if $|x| > 1$,

$$y = x + \frac{1}{2} + \frac{3}{8x} + 0\left(\frac{1}{x^2}\right)$$

therefore $y = x + \dfrac{1}{2}$ is an asymptote approached from above as $x \to \infty$ and from below as $x \to -\infty$.

As $x \to 1+$, $y \to \infty$, therefore $x = 1$ is an asymptote. Near the origin, $y \simeq -|x|^{3/2} \to 0-$. Turning points are where

$$\frac{(x - 3/2)\sqrt{x}}{(x-1)^{3/2}} = 0$$

hence at $(0, 0)$ or $(3/2, 3\sqrt{3}/2)$. See Figure 20 where the asymptotes are shown by dashed lines.

(ii)   There is symmetry about the line $y = x$. There is only one intercept with the axes, at the origin. Substituting $y = tx$ in the equation yields $x^3(1 + t^3) = 3tx^2$ from which we deduce that $x = 3t/(1 + t^3)$ and hence $y = 3t^2/(1 + t^3)$.

At $t = 0$, $(x, y) = (0, 0)$; as $t \to \infty$ $(x, y) \to (0, 0)$ again.

At $t = 1$, $(x, y) = \left(\dfrac{3}{2}, \dfrac{3}{2}\right)$

At $t = \dfrac{1}{2}$, $(x, y) = \left(\dfrac{4}{3}, \dfrac{2}{3}\right)$

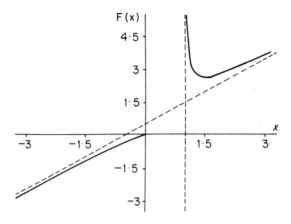

Figure 20

At $t = 2$, $(x, y) = \left(\dfrac{2}{3}, \dfrac{4}{3}\right)$ etc.

Clearly, $t = -1$ is a crucial value.

As $t \to -1$ from above, $x \to -\infty$, $y \to +\infty$
As $t \to -1$ from below, $x \to +\infty$, $y \to -\infty$.
As $t \to -\infty$, $(x, y) \to (0, 0)$ from negative y and positive x.
As $t \to +\infty$, $(x, y) \to (0, 0)$ from positive y and x.

Note further that

$$x + y = \frac{3t(1 + t)}{1 + t^3} = \frac{3t}{1 - t + t^2}$$

As $t \to -1$, $x + y \to -1$, and we conclude that $x + y + 1 = 0$ is an asymptote. Finally,

$$\frac{dy}{dx} = \frac{dy}{dt} \div \frac{dx}{dt} = \frac{3t(2 - t^3)}{(1 + t^3)^2} \div \frac{3(1 - 2t^3)}{(1 + t^3)^2} = \frac{t(2 - t^3)}{(1 - 2t^3)}$$

so that horizontal gradients occur when $t = 0$ or $\sqrt[3]{2}$ corresponding to the points $(x, y) = (0, 0)$ and $(\sqrt[3]{2}, \sqrt[3]{4})$, i.e. $(0, 0)$ and approximately $(1.26, 1.59)$.

A vertical gradient occurs when $t^3 = \frac{1}{2}$ giving $(x, y) = (\sqrt[3]{4}, \sqrt[3]{2})$, which we could have deduced from symmetry. This suggests we might also expect a vertical gradient at $(0, 0)$ and this corresponds

40

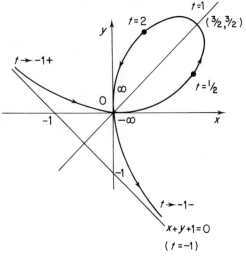

$t=1$

$t=2$

$(\sqrt[3]{2},\sqrt[3]{2})$

y

$t \rightarrow -1+$

∞

O

$t=\tfrac{1}{2}$

$-\infty$

-1

x

-1

$t \rightarrow -1-$

$x+y+1=0$

$(t=-1)$

Figure 21

to $t = \infty$ which can be seen to be so from the formula for dy/dx. Pooling the information so far we produce Figure 21 in which the asymptote $x + y + 1 = 0$ and the line of symmetry $y = x$ are shown.

Example 20

(a) Sketch the curve $(1 - x)^2(2 + y)^3 = 1$ showing features such as maxima, minima and asymptotes.

(b) For another curve whose equation is $(1 - x^2)^2(2 + y)^3 = 1$ show that

$$\frac{dy}{dx} = \frac{4x}{3(1 - x^2)^{5/3}}$$

and hence deduce the nature of the curve $y(x)$ in the neighbourhood of the lines $x = 0$, $x = \pm 1$ and $y = -2$.   (CEI)

(a) Symmetry about $x = 1$. (Put $u = 1 - x$ and replace u by $-u$.) The intecepts with the axes are at $x = 1 \pm (1/2\sqrt{2})$ and at $y = -1$. It is

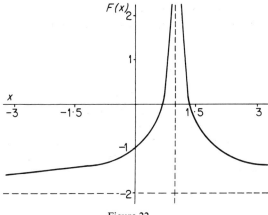

Figure 22

necessary for $y > -2$ in order that real values of x come from the equation of the curve shown in Figure 22.

As $y \to \infty$, $x^2 \to 1$; so that $x \to 1+$ or $1-$
As $x \to \pm\infty$, $x^2 \to \infty$; so that $y \to -2+$ (remember $y > -2$)

Hence $x = 1$ and $y = -2$ are asymptotes. There are no turning points, since $y' = 2/3(1 - x)^{1/3}$.

(b) This curve, shown in Figure 23, is one with an equation similar to

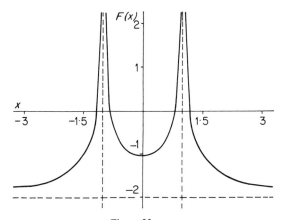

Figure 23

that in (a) where x has been replaced by x^2. This time there *is* symmetry about the y-axis. The intercepts with the axes are at $y = -1$ and at

$$x = \pm \sqrt{\left(1 \pm \frac{1}{2\sqrt{2}}\right)}$$

i.e. $x = \pm 0.804$, ± 1.163. Note that near $x = 0$, dy/dx has the sign of x itself. Hence there is a local minimum at $x = 0$ of -1. Near $x = +1$, $dy/dx \to +\infty$ as $x \to 1+$ and $x \to 1-$. Hence $x = 1$ is an asymptote. Similar arguments show that $x = -1$ and $y = -2$ are asymptotes. (We would have expected an asymptote of $x = -1$ by symmetry.)

E

Differentiation and Applications

Example 21

(a) Show that $y = \operatorname{cosec} x / \sinh(1/x)$ satisfies the relation

$$\frac{\mathrm{d}y}{\mathrm{d}x} = y\left[-\cot x + \frac{1}{x^2}\coth\left(\frac{1}{x}\right)\right]$$

(b) In Figure 24 AXB represents a road connecting towns A and B. In suitable units, the cost per unit length of laying the section AX is C_1 and that of XB is C_2, where $C_2 \geqslant C_1$.

Denoting the total cost of laying AXB as C, show by investigating $\mathrm{d}C/\mathrm{d}x$ that C will be minimized for some value of x in the range $0 < x < d$ (i.e. for some point X between P and Q) and find a quartic equation satisfied by this value. In particular, if $C_1 = C_2$, show that

$$x = \frac{l_1 d}{l_1 + l_2}$$

What can you therefore say about lines AX and XB?

(CEI)

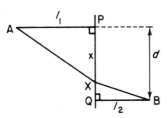

Figure 24

43

(a) Using the quotient rule,

$$y' = \frac{\sinh\left(\frac{1}{x}\right)[-\operatorname{cosec} x . \cot x] - \operatorname{cosec} x\left[\left(-\frac{1}{x^2}\right)\cosh\left(\frac{1}{x}\right)\right]}{\sinh^2\left(\frac{1}{x}\right)}$$

$$= \frac{\operatorname{cosec} x}{\sinh\left(\frac{1}{x}\right)}(-\cot x) + \frac{\operatorname{cosec} x}{\sinh\left(\frac{1}{x}\right)}\left[\frac{1}{x^2}\coth\left(\frac{1}{x}\right)\right]$$

$$= y\left[-\cot x + \frac{1}{x^2}\coth\left(\frac{1}{x}\right)\right], \text{ as required.}$$

(b) $C = C_1\left(l_1^2 + x^2\right)^{1/2} + C_2\left(l_2^2 + (d-x)^2\right)^{1/2}$

therefore

$$\frac{\mathrm{d}C}{\mathrm{d}x} = \frac{C_1 . x}{\left(l_1^2 + x^2\right)^{1/2}} - \frac{C_2(d-x)}{\left(l_2^2 + (d-x)^2\right)^{1/2}}$$

This is zero when

$$C_1^2 x^2\left(l_2^2 + (d-x)^2\right) - C_2^2(d-x)^2\left(l_1^2 + x^2\right) = 0$$

which is the required quartic equation.

It can be shown that $\mathrm{d}^2 C/(\mathrm{d}x^2) > 0$ for all x, so that the stationary point will be a minimum.

If $C_1 = C_2$, the quartic equation reduces to

$$x^2\left(l_2^2 + (d-x)^2\right) = (d-x)^2\left(l_1^2 + x^2\right)$$

or $x^2 l_2^2 = (d-x)^2 l_1^2$ so that

$$x^2(l_2^2 - l_1^2) + 2dl_1^2 x - d^2 l_1^2 = 0.$$

Hence

$$x = \frac{-dl_1^2 \pm \sqrt{d^2 l_1^4 + d^2 l_1^2(l_2^2 - l_1^2)}}{l_2^2 - l_1^2};$$

therefore

$$x = \frac{-dl_1^2 \pm dl_1 l_2}{l_2^2 - l_1^2}$$

giving

$$x = \frac{-\mathrm{d}l_1(l_1 + l_2)}{l_2{}^2 - l_1{}^2} \quad \text{or} \quad \frac{\mathrm{d}l_1(l_2 - l_1)}{l_2{}^2 - l_1{}^2}$$

i.e.

$$x = \frac{-\mathrm{d}l_1}{l_2 - l_1} \quad \text{or} \quad \frac{\mathrm{d}l_1}{l_2 + l_1}$$

It can be shown that the first root cannot lie in $0 < x < d$. Similar triangles shows that AXB will be a straight line. Is this an expected result?

Example 22

(a) Determine the first three non-vanishing terms of the Maclaurin series for sec x.
Hence evaluate

$$\lim_{x \to 0} \frac{\sec x - 1}{x \sin x}$$

use L'Hôpital's rule to verify this result. (CEI)

(b) State L'Hôpital's rule for evaluating

$$\lim_{t \to 0} \frac{f(t)}{g(t)}.$$

The response $y(t)$ of a certain hydraulic valve subject to sinusoidal input variation is given by

$$\frac{\mathrm{d}y}{\mathrm{d}t} = \sqrt{2\left(1 - \frac{y^2}{\sin^2 t}\right)}$$

where $y = 0$ at $t = 0$.
Show that

$$\lim_{t \to 0} \left(\frac{\mathrm{d}y}{\mathrm{d}t}\right) = \sqrt{\frac{2}{3}} \qquad \text{(CEI)}$$

(a) If $y = \sec x$, then

$$y' = \sec x \tan x, \qquad y'' = \sec x \tan^2 x + \sec^3 x,$$
$$= 2\sec^3 x - \sec x = 2y^3 - y$$

therefore

$$y''' = 6y^2 y' - y', \qquad y^{iv} = 12y.(y')^2 + 6y^2 y'' - y''$$

Hence at $x = 0$,

$$y = 1, \qquad y' = 0, \qquad y'' = 1, \qquad y''' = 0, \qquad y^{iv} = 5$$

and

$$\sec x = 1 + \frac{x^2}{2} + \frac{5}{24} x^4 + \dots$$

Using this result and the standard series for $\sin x$,

$$\frac{\sec x - 1}{x \sin x} = \frac{\frac{1}{2} x^2 + 0(x^4)}{x^2 + 0(x^4)} = \frac{\frac{1}{2} + 0(x^2)}{1 + 0(x^2)} \to \frac{1}{2} \text{ as } x \to 0$$

Alternatively, using L'Hôpital's rule twice,

$$\lim_{x \to 0} \left(\frac{\sec x - 1}{x \sin x} \right) = \lim_{x \to 0} \left(\frac{\sec x \tan x}{\sin x + x \cos x} \right)$$

$$= \lim_{x \to 0} \left(\frac{2 \sec^3 x - \sec x}{2 \cos x - x \sin x} \right) = \frac{1}{2}$$

(b) If $f(a) = g(a) = 0$ but $g'(a) \neq 0$, and if both limits exist then

$$\lim_{t \to a} \left\{ \frac{f(t)}{g(t)} \right\} = \lim_{t \to a} \left\{ \frac{f'(t)}{g'(t)} \right\}.$$

Let

$$\lim_{t \to 0} \left(\frac{dy}{dt} \right) = \dot{y}_0$$

Then, taking the limit as $t \to 0$ of the given differential equation,

$$\dot{y}_0 = \sqrt{2 \left\{ 1 - \lim_{t \to 0} \left(\frac{y^2}{\sin^2 t} \right) \right\}} \qquad (1)$$

using the rules for limits of sums, differences, products etc.

From the initial condition it follows that

$$\frac{y}{\sin t} \to \frac{0}{0}$$

Using L'Hôpital's rule,

$$\lim_{t \to 0} \left(\frac{y}{\sin t} \right) = \lim_{t \to 0} \left(\frac{\dot{y}}{\cos t} \right) = \frac{\dot{y}_0}{1}$$

Therefore from (1) $\dot{y}_0 = \sqrt{2(1 - \dot{y}_0^2)}$,

from which we obtain $\underline{\dot{y}_0 = \sqrt{\tfrac{2}{3}}}$.

Example 23

(a) If $y = \ln(\sec x + \tan x) - (2 - \sec x)\tan x$, show that $(dy/dx) \geqslant 0$ over the range $0 \leqslant x \leqslant (\pi/2)$, and hence that y is also non-negative.

(b) If $y = \ln \cos x$, prove that

$$\frac{d^3 y}{dx^3} + 2\frac{d^2 y}{dx^2} \cdot \frac{dy}{dx} = 0.$$

Hence, or otherwise, obtain the Maclaurin expansion of y as far as the term in x^5.

Using the substitution $x = \pi/4$, deduce the approximation

$$\ln 2 \simeq \frac{\pi^2}{16}\left(1 + \frac{\pi^2}{96}\right).$$

Finally, evaluate

$$\lim_{x \to 0}\left(\frac{\ln \cos x}{1 - \cos x}\right)$$

(CEI)

(Note that $\log_e x$ and $\ln x$ are equivalent.)

(a) $\dfrac{dy}{dx} = \dfrac{\sec x \tan x + \sec^2 x}{\sec x + \tan x} - 2 \sec^2 x + \sec x \tan x \tan x + \sec^3 x$

$\quad = \sec x - 2 \sec^2 x + \sec x(\sec^2 x - 1) + \sec^3 x$

$\quad = 2 \sec^2 x(\sec x - 1)$

Hence

$$\frac{dy}{dx} \geqslant 0 \quad \text{for} \quad 0 \leqslant x \leqslant \frac{\pi}{2}.$$

Since

$$y = 0 \quad \text{at} \quad x = 0, \qquad y \geqslant 0 \quad \text{for} \quad 0 \leqslant x \leqslant \frac{\pi}{2}.$$

(b) By successive differentiation we obtain

$$y' = -\tan x, \qquad y'' = -\sec^2 x, \qquad \begin{aligned} y''' &= -2\sec^2 x . \tan x \\ &= -2y'' . y' \end{aligned}$$

Hence

$$y''' + 2y'' . y' = 0 \qquad (1)$$

Two further differentiations produce

$$y^{iv} + 2y''' . y' + 2(y'')^2 = 0 \qquad (2)$$

and

$$y^{v} + 2y^{iv}. y' + 6y''' . y'' = 0 \qquad (3)$$

The values for y, y', and y'' at $x = 0$ are found by direct evaluation to be $0, 0, -1$, respectively. The values of y''', y^{iv} and y^{v} at $x = 0$ are found from (1), (2), and (3), respectively to be $0, -2, 0$.

Hence

$$y = -\tfrac{1}{2} x^2 - \tfrac{1}{12} x^4 + 0(x^6) \qquad (4)$$

At $x = \pi/4$, using (4), $y \simeq -\dfrac{1}{2}\left(\dfrac{\pi}{4}\right)^2 - \dfrac{1}{12}\left(\dfrac{\pi}{4}\right)^4$

$$= -\frac{\pi^2}{32}\left(1 + \frac{\pi^2}{96}\right)$$

But,

$$y = \log_e \cos \frac{\pi}{4} = \log_e\left(\frac{1}{\sqrt{2}}\right) = -\frac{1}{2}\log_e 2.$$

Hence the result.

Finally, using standard series,

$$\frac{\log_e \cos x}{1 - \cos x} = \frac{-\tfrac{1}{2} x^2 + 0(x^4)}{\tfrac{1}{2} x^2 + 0(x^4)}$$

and hence the limit as $x \to 0$ is -1. Alternatively, from L'Hôpital's rule applied twice

$$\lim_{x \to 0}\left(\frac{\log_e \cos x}{1 - \cos x}\right) = \lim_{x \to 0}\left(\frac{-\tan x}{\sin x}\right) = \lim_{x \to 0}\left(\frac{-\sec^2 x}{\cos x}\right) = -1$$

Example 24

(a) Use Taylor's expansion to show that:

$$y = \frac{\log_e(2-x)}{x^3 - 3x + 2} \quad \text{behaves like} \quad -\frac{1}{3}\frac{1}{(x-1)} \quad \text{for } x \text{ near } 1$$
(LUT)

(b) A column of length l has a vertical load P and a horizontal load F at the top. The transverse deflection is

$$\delta = \frac{Fl}{P}\left[\frac{\tan ml}{ml} - 1\right]$$

where $m^2 = P/EI$. Show that as $P \to 0$, $\delta \to Fl^3/3EI$ and that a small value of P leads to an increase of this by about $(40\, Pl^2/EI)\%$. (Assume that $\tan x = x + \frac{1}{3}x^3 + \frac{2}{15}x^5 + \ldots$)
(LUT)

(a) The first thing to notice is that the expression $-\frac{1}{3}[1/(x-1)]$ misbehaves at $x = 1$; a sketch of the curve (Figure 25) shows how. Hence, this is the behaviour we expect the given function to follow near $x = 1$.

Put $u = x - 1$ so that x near 1 is equivalent to u near zero. Then $2 - x = 1 - u$ so that a Maclaurin expansion gives

$$\log_e(2 - x) = \log_e(1 - u) = -u - \frac{u^2}{2} - \frac{u^3}{3} - \ldots$$

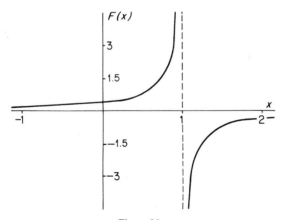

Figure 25

Also

$$x^3 - 3x + 2 = (x-1)^2(x+2) = u^2(u+3)$$

Therefore,

$$y = \frac{-u(1 - u/2 - u^2/3 - \ldots)}{u^2(u+3)}$$

$$= -\frac{1}{u}\left(1 - \frac{u}{2} - \frac{u^2}{3} - \cdots\right) \cdot \frac{1}{u+3}$$

Near $u = 0$, $u + 3 \simeq 3$ and the expression in round brackets $\simeq 1$ so that

$$y \simeq -\frac{1}{3u} = -\frac{1}{3} \cdot \frac{1}{x-1},$$

as required.

(b)

$$\frac{\tan ml}{ml} - 1 = \frac{ml + \frac{1}{3}m^3l^3 + \frac{2}{15}m^5l^5 + \cdots}{ml} - 1$$

$$\simeq \frac{1}{3}m^2l^2 + \frac{2}{15}m^4l^4$$

therefore

$$\delta \simeq \frac{Fl}{P}\left[\frac{1}{3}\frac{Pl^2}{EI} + \frac{2}{15}\frac{P^2l^4}{(EI)^2}\right]$$

$$= \frac{Fl^3}{3EI} + \frac{2}{15}\frac{Pl^5F}{(EI)^2}$$

Hence, as $P \to 0$

$$\delta \to \underline{\frac{Fl^3}{3EI}}$$

The approximate percentage increase required is

$$\left(\frac{2}{15}\frac{Pl^5F}{(EI)^2} \times \frac{3EI}{Fl^3} \times 100\right) \%$$

$$= \left(\frac{2}{5} \cdot \frac{Pl^2}{EI} \cdot 100\right) \%$$

$$= \underline{\left(\frac{40Pl^2}{EI}\right) \%}$$

Example 25

(a) Find the stationary points of the curve

$$y = x^2 \exp(-x)$$

Indicate on a sketch the important features, marking numerical values.

Consider the family of curves $y = x^n \exp(-x)$, where $n \geqslant 2$ is an integer. Explain the changes in the curves as n increases.

(CEI)

(b) The efficiency of a certain thermal cycle is

$$\eta = 1 - \frac{T_1}{T_2 - T_1} \log_e\left(\frac{T_2}{T_1}\right)$$

where T_1 and $T_2 \geqslant T_1$ are temperatures at specific stages of the process.

Assuming that $T_2 = T_1(1 + x)$, obtain η as a function of the non-dimensional variable x and deduce the value of $\underset{x \to 0}{\text{Lim}}\, \eta$, using (i) a Maclaurin series and (ii) L'Hôpital's rule. (CEI)

(a) $y = x^2\,e^{-x}$ therefore $dy/dx = (2x - x^2)\,e^{-x}$.

At a stationary point $2x - x^2 = 0$, i.e. $x = 0$ or $x = 2$.

$$\frac{d^2 y}{dx^2} = (2 - 4x + x^2)\,e^{-x}$$

so that $y(0) = 0$ is a local minimum and $y(2) = 4e^{-2}$ is a local maximum.

As $x \to \infty$, $y \to 0+$ and as $x \to -\infty$, $y \to \infty$

Any points of inflexion occur where $2 - 4x + x^2 = 0$, i.e. at $x \simeq 0.6$ and 3.4. Refer to Figure 26(a).

In general, if $y = x^n\,e^{-x}$ then

$$\frac{dy}{dx} = (n - x)x^{n-1}\,e^{-x} \quad \text{and} \quad \frac{d^2 y}{dx^2} = \left[(n^2 - n) - 2nx + x^2\right]x^{n-2}\,e^{-x}$$

There is always a local minimum at $x = 0$ of 0 and a local maximum at $x = n$ of $n^n e^{-n}$.

As n increases, so does $n^n e^{-n}$;

$$\left\{\frac{(n+1)^{n+1}e^{-n-1}}{n^n e^{-n}} = \left(1 + \frac{1}{n}\right)^n (n+1)e^{-1}\right\} > 1$$

As n increases the peaks move in the positive x-direction and increase in magnitude; see Figure 26(b).

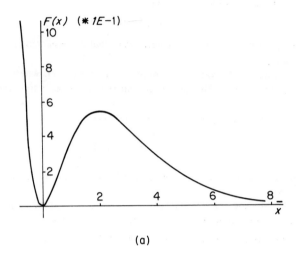

(a)

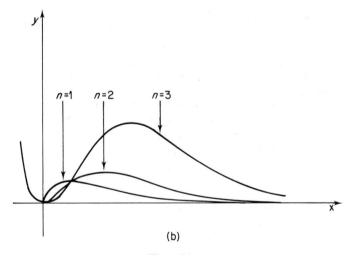

(b)

Figure 26

Maclaurin's series for $\log_e(1 + x)$, valid for $-1 < x \leqslant 1$, is

$$\log_e(1 + x) = x - \tfrac{1}{2}x^2 + \tfrac{1}{3}x^3 - \ldots$$

Using the given substitution for T_2,

$$\eta = 1 - \frac{1}{x}\log_e(1 + x)$$

(i)

$$\eta = \tfrac{1}{2}x - \tfrac{1}{3}x^2 + \ldots$$
$$\to 0 \text{ as } x \to 0$$

(ii)   Since $\dfrac{1}{x}\log_e(1 + x) \to \dfrac{0}{0}$ as $x \to 0$, we can use L'Hôpital's
rule:

$$\lim_{x \to 0} \eta = 1 - \lim_{x \to 0}\left\{\frac{\dfrac{1}{x + 1}}{1}\right\} = 1 - 1 = 0, \text{ again.}$$

F

Partial Differentiation

Example   26

In Figure 27 QPR represents a ray of light reflected at P in a cylindrical mirror, centre O, where QP is parallel to Ox.

Obtain the equation of PR in the form $f(x, y, \theta) = 0$. Hence from the equations

$$f(x, y, \theta) = 0, \qquad \frac{\partial f}{\partial \theta} = 0$$

show that the envelope of the reflected ray PR as θ varies (the caustic curve) is given by

$$x = \tfrac{1}{2}\, a(3 - 2 \cos^2\theta) \cos\theta, \qquad y = a \sin^3\theta$$

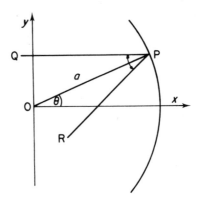

Figure 27

Finally, deduce the result

$$\frac{dy}{dx} = \tan 2\theta$$

and find an expression for

$$\frac{d^2y}{dx^2} \tag{CEI}$$

P has coordinates ($a \cos \theta$, $a \sin \theta$). From the geometry of the diagram $Q\hat{P}O = \theta$, and by the law of reflection $O\hat{P}R = \theta$. Hence RP makes an angle 2θ with the x-axis. The equation of RP is

$$y - a \sin \theta = \tan 2\theta (x - a \cos \theta)$$

or

$$f(x, y, \theta) \equiv y - a \sin \theta - \tan 2\theta (x - a \cos \theta) = 0 \tag{1}$$

Now

$$\frac{\partial f}{\partial \theta} \equiv -a \cos \theta - 2 \sec^2 2\theta . x + 2 \sec^2 2\theta . a \cos \theta$$

$$-a \tan 2\theta . \sin \theta = 0$$

so that

$$2 \sec^2 2\theta . x = 2 \sec^2 2\theta . a \cos \theta - a \left[\frac{\cos \theta \cos 2\theta + \sin 2\theta \sin \theta}{\cos 2\theta} \right]$$

$$= 2 \sec^2 2\theta . a \cos \theta - a \cos \theta . \sec 2\theta$$

This last equation can be reduced to

$$x = \tfrac{1}{2} a \cos \theta (2 - \cos 2\theta) = \tfrac{1}{2} a \cos \theta (3 - 2 \cos^2\theta) \tag{2}$$

To find the envelope of PR solve (1) and (2) simultaneously. Substitute (2) in (1) to produce

$$y = a \sin \theta + \tan 2\theta (\tfrac{1}{2} a \cos \theta [1 - 2 \cos^2 \theta])$$

i.e.

$$y = a \sin \theta - 2 \sin \theta . \cos \theta . \tfrac{1}{2} a \cos \theta$$

i.e.

$$y = a \sin^3 \theta \tag{3}$$

The parametric equations of the envelope are (2) and (3). Since

$$\frac{\mathrm{d}y}{\mathrm{d}x} = \frac{\mathrm{d}y}{\mathrm{d}\theta} \div \frac{\mathrm{d}x}{\mathrm{d}\theta},$$

it follows that

$$\frac{\mathrm{d}y}{\mathrm{d}x} = \frac{3\,a\,\sin^2\theta\,.\,\cos\theta}{\frac{1}{2}a(-3\sin\theta + 6\cos^2\theta\,\sin\theta)} = \frac{3\,a\,\sin\theta\,.\,\frac{1}{2}\sin 2\theta}{\frac{3}{2}a\,\sin\theta(-1 + 2\cos^2\theta)}$$

$$= \frac{\frac{3}{2}a\,\sin\theta\,\sin 2\theta}{\frac{3}{2}a\,\sin\theta\,\cos 2\theta}$$

$$= \underline{\tan 2\theta}$$

We must now beware:

$$\frac{\mathrm{d}^2 y}{\mathrm{d}x^2} = \frac{\mathrm{d}}{\mathrm{d}x}\left(\frac{\mathrm{d}y}{\mathrm{d}x}\right) = \frac{\mathrm{d}}{\mathrm{d}\theta}\left(\frac{\mathrm{d}y}{\mathrm{d}x}\right) \div \frac{\mathrm{d}x}{\mathrm{d}\theta}$$

$$= \frac{2\sec^2 2\theta}{\frac{3}{2}a\,\sin\theta\,.\,\cos 2\theta}$$

$$= \underline{\frac{4}{3}\sec^3 2\theta\,\operatorname{cosec}\theta}$$

Example 27

(a) Find the location and nature of the stationary points of

$$z = xy(4x + 2y + 1).$$

(b) A rectangular sheet of metal of width $2l$ is bent to form a trough without ends. The cross-section is a polygon ABCDE as shown below in Figure 28. Prove that as x and θ vary, the maximum value of the cross-sectional area is $l^2/\sqrt{3}$. (LUT)

(a) $$z = 4x^2 y + 2xy^2 + xy$$

therefore

$$z_x = 8xy + 2y^2 + y; \qquad z_y = 4x^2 + 4xy + x;$$
$$z_{xx} = 8y; \qquad z_{xy} = 8x + 4y + 1 = z_{yx}; \qquad z_{yy} = 4x.$$

Location $z_x = 0 = z_y$, i.e.

$$y(8x + 2y + 1) = 0 \qquad \text{(i)}$$
$$x(4x + 4y + 1) = 0 \qquad \text{(ii)}$$

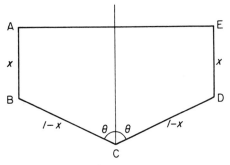

Figure 28

Each equation has two possible solutions, leading to four possible combinations. These can be solved to give the points $(0,0)$, $(0, -\frac{1}{2})$, $(-\frac{1}{4}, 0)$, $(-\frac{1}{12}, -\frac{1}{6})$.

Nature $D \equiv z_{xx} \cdot z_{xy} - (z_{xy})^2$. Refer to Table 2.

Note that if $D > 0$ then z_{xx} and z_{yy} must have the same sign and we examine z_{xx} by choice.

Table 2

Point	$(0,0)$	$(0, -\frac{1}{2})$	$(-\frac{1}{4}, 0)$	$(-\frac{1}{12}, -\frac{1}{6})$
z_{xx}	0	-4	0	$-4/3$
z_{yy}	0	0	-1	$-1/3$
z_{xy}	1	-1	-1	$-1/3$
D	<0	<0	<0	>0
	Saddle point	Saddle point	Saddle point	Local maximum (since $z_{xx} < 0$).

(b)   Area of cross-section is

$$A = 2x(l - x)\sin \theta + 2 \cdot \tfrac{1}{2} \cdot (l - x)^2 \sin \theta \cos \theta$$

therefore

$$\frac{\partial A}{\partial x} = 2(l - 2x)\sin \theta - 2(l - x)\sin \theta \cdot \cos \theta$$

58

and

$$\frac{\partial A}{\partial \theta} = 2x(l-x)\cos\theta + (l-x)^2(\cos^2\theta - \sin^2\theta)$$

Since $l \neq x$ and $\sin\theta \neq 0$ the equations

$$\frac{\partial A}{\partial x} = 0 \quad \text{and} \quad \frac{\partial A}{\partial \theta} = 0$$

reduce to

$$\cos\theta = \frac{(l-2x)}{(l-x)}, \qquad 2x\cos\theta + (l-x)(2\cos^2\theta - 1) = 0$$

Substitution of the first equation into the second produces

$$3x^2 - 4xl + l^2 = 0$$

from which it follows that $x = l/3$, $\cos\theta = \frac{1}{2}$ and the value of A, which is clearly a maximum (since the minimum value is 0), is $\underline{l^2/\sqrt{3}}$.

Example 28

(a) The diameter of the base of a right circular cone and the slant height are measured as 6 cm and 5 cm respectively. Each measurement is subject to a maximum error of ± 0.1 cm. Find the greatest error in the calculated value of (i) the total surface area, (ii) the volume of the cone. What are the corresponding relative errors?

(b) The area S of a triangle ABC is obtained from measurements of sides b and c and angle A. Show that the error δS due to small errors δb, δc and δA is given by

$$\delta S/S \simeq \delta b/b + \delta c/c + \cot A \cdot \delta A$$

Interpret this result when $A = \pi/2$.

(c) Find a formula for p in terms of c, q and k such that the function $\phi = A\exp(-kt/2)\sin pt \cdot \cos qx$ satisfies the equation

$$\frac{\partial^2\phi}{\partial x^2} = \frac{1}{c^2}\left[\frac{\partial^2\phi}{\partial t^2} + k\frac{\partial\phi}{\partial t}\right]$$

(CEI)

(a) For a cone of slant height l and base diameter d,

Surface area $S = \frac{1}{2}\pi dl + \frac{1}{4}\pi d^2$

Volume $\qquad V = \frac{1}{3}\pi \frac{d^2}{4}\sqrt{l^2 - \frac{1}{4}d^2}$

$\qquad\qquad\quad = \frac{1}{24}\pi d^2 \sqrt{4l^2 - d^2}$

(i) $\qquad\qquad \delta S \simeq \frac{\partial S}{\partial d}\delta d + \frac{\partial S}{\partial l}\delta l$, since errors are small.

$\qquad\qquad\quad = \frac{1}{2}\pi(l + d)\,\delta d + \frac{1}{2}\pi d\,\delta l$

$\qquad\qquad\quad = \frac{11}{2}\pi\,\delta d + 3\pi\,\delta l$

therefore

$$|\delta S| \leqslant \tfrac{11}{2}\pi(0.1) + 3\pi(0.1) = 0.85\pi = \underline{2.67\ \text{cm}^2}$$

(ii) $\quad \delta V \simeq \frac{\pi}{24}\left[2d\sqrt{4l^2 - d^2} - \frac{d^3}{\sqrt{4l^2 - d^2}}\right]\delta d + \left[\frac{1}{6}\frac{\pi d^2 l}{\sqrt{4l^2 - d^2}}\right]\delta l$

$\qquad\qquad = \frac{69}{24}\pi\,\delta d + \frac{15}{4}\pi\,\delta l$

therefore

$$|\delta V| \leqslant \frac{69}{24}\pi(0.1) + \frac{15}{4}\pi(0.1) = \frac{15.9}{24}\pi = \underline{2.08\ \text{cm}^3}$$

Calculated values are $S = 24\pi$, $V = 12\pi$ so that maximum relative errors are:

$$\left|\frac{\delta S}{S}\right| \leqslant 0.0354 \simeq 3\tfrac{1}{2}\%\ \text{in } S$$

$$\left|\frac{\delta V}{V}\right| \leqslant 0.0552 \simeq 5\tfrac{1}{2}\%\ \text{in } V$$

(b) $S = \frac{1}{2}bc \sin A$

$$\delta S \simeq \frac{\partial S}{\partial b}\cdot\delta b + \frac{\partial S}{\partial c}\cdot\delta c + \frac{\partial S}{\partial A}\cdot\delta A$$

$$= \tfrac{1}{2}c \sin A\,\delta b + \tfrac{1}{2}b \sin A\,\delta c + \tfrac{1}{2}bc \cos A\,\delta A$$

therefore

$$\frac{\delta S}{S} \simeq \frac{\frac{1}{2}c \sin A}{\frac{1}{2}bc \sin A} \delta b + \frac{\frac{1}{2}b \sin A}{\frac{1}{2}bc \sin A} \delta c + \frac{\frac{1}{2}bc \cos A}{\frac{1}{2}bc \sin A} \delta A$$

$$= \frac{\delta b}{b} + \frac{\delta c}{c} + \cot A \, \delta A,$$

as required.

When $A = \pi/2$, $\cot A = 0$ so that

$$\frac{\delta S}{S} \simeq \frac{\delta b}{b} + \frac{\delta c}{c}$$

and in this case the error in S is approximately independent of any error in A.

(c)
$$\frac{\partial \phi}{\partial x} = - Aq e^{-kt/2} \sin pt \cos qx, \quad \frac{\partial^2 \phi}{\partial x^2} = - q^2 \phi$$

$$\frac{\partial \phi}{\partial t} = A e^{-kt/2}(- \tfrac{1}{2}k \sin pt + p \cos pt). \cos qx$$

$$\frac{\partial^2 \phi}{\partial t^2} = A e^{-kt/2}(\tfrac{1}{4}k^2 \sin pt - \tfrac{1}{2}kp \cos pt$$

$$- \tfrac{1}{2}kp \cos pt - p^2 \sin pt)\cos qx$$

Hence, substitution into the given equation gives,

$$- q^2\phi = \frac{1}{c^2}[A e^{-kt/2}\{(- p^2 - \tfrac{1}{4}k^2)\sin pt + (0)\cos pt\}\cos qx]$$

$$= - \frac{1}{c^2}(p^2 + \tfrac{1}{4}k^2)\phi$$

so that

$$p^2 = c^2q^2 - \tfrac{1}{4}k^2.$$

Example 29

(a) On the pressure—volume diagram for a fluid one particular curve of constant temperature always has a point of inflexion, together with a horizontal tangent; this point is called the *critical* point.

A van der Waals' gas obeys the law

$$p = \frac{RT}{V-b} - \frac{a}{V^2}$$

where p is pressure, V is specific volume, T is temperature, and a and b are constants.

Show that at the critical point for this gas

$$V = 3b \quad \text{and} \quad \frac{pV}{RT} = \frac{3}{8}.$$

(CEI)

(b) If

$$U = \frac{x^2 y^2}{x^2 + y^2},$$

show that

(i) $x \dfrac{\partial U}{\partial x} + y \dfrac{\partial U}{\partial y} = 2U,$

and

(ii) $x^2 \dfrac{\partial^2 U}{\partial x^2} + 2xy \dfrac{\partial^2 U}{\partial x \, \partial y} + y^2 \dfrac{\partial^2 U}{\partial y^2} = 2U.$

(CEI)

(c) A point is moving on the surface $z = (x^3 - 4y^2)\exp\{-(x^2 + y^2)\}$ in the plane $y = x$.

Obtain the rate of change of z if x is increasing at the rate of 3 units/second when

(i) $x = 1$ (ii) x is large.

(CEI)

(a) With T constant, we find

$$\frac{\partial p}{\partial V} = -\frac{RT}{(V-b)^2} + \frac{2a}{V^3} \quad \text{and} \quad \frac{\partial^2 p}{\partial V^2} = \frac{2RT}{(V-b)^3} - \frac{6a}{V^4}.$$

At the critical point, therefore,

$$\frac{RT}{(V-b)^2} = \frac{2a}{V^3} \quad \text{(i)} \quad \text{and} \quad \frac{2RT}{(V-b)^3} = \frac{6a}{V^4} \quad \text{(ii)}$$

so that, substituting (i) into (ii)

$$\frac{2}{(V-b)} \frac{2a}{V^3} = \frac{6a}{V^4}$$

from which

$$2V = 3(V - b)$$

and

$$\underline{V = 3b} \qquad \text{(iii)}$$

Now

$$RT = \frac{2a(V-b)^2}{V^3} \qquad \text{from (i)}$$

and at the critical point

$$RT = \frac{2a \cdot (3b - b)^2}{27b^3} = \frac{8a}{27b}, \qquad \text{using (iii)}$$

so that

$$p = \frac{8a}{27b \cdot 2b} - \frac{a}{9b^2} = \frac{a}{27b^2}$$

and

$$\frac{pV}{RT} = \frac{a}{27b^2} \cdot 3b \cdot \frac{27b}{8a} = \underline{\frac{3}{8}}.$$

(b) (i) $$\frac{\partial U}{\partial x} = \frac{(x^2 + y^2)2xy^2 - x^2y^2 \cdot 2x}{(x^2 + y^2)^2}$$

$$= \frac{2xy^4}{(x^2 + y^2)^2}$$

and, by symmetry,

$$\frac{\partial U}{\partial y} = \frac{2x^4y}{(x^2 + y^2)^2}$$

Therefore,

$$x\frac{\partial U}{\partial x} + y\frac{\partial U}{\partial y} = \frac{2x^2y^2}{(x^2 + y^2)^2}(y^2 + x^2) = 2U.$$

(ii) From (i) $$\left(x\frac{\partial}{\partial x} + y\frac{\partial}{\partial y}\right)U = 2U$$

so that

$$\left(x\frac{\partial}{\partial x} + y\frac{\partial}{\partial y}\right)^2 U = \left(x\frac{\partial}{\partial x} + y\frac{\partial}{\partial y}\right)2U = 4U \qquad \text{(from (i))}$$

Now

$$\left(x\frac{\partial}{\partial x} + y\frac{\partial}{\partial y}\right)^2 U \equiv \left(x\frac{\partial}{\partial x} + y\frac{\partial}{\partial y}\right)\left(x\frac{\partial U}{\partial x} + y\frac{\partial U}{\partial y}\right)$$

$$\equiv x^2\frac{\partial^2 U}{\partial x^2} + x.1.\frac{\partial U}{\partial x} + xy\frac{\partial^2 U}{\partial x\,\partial y}$$

$$+ yx\frac{\partial^2 U}{\partial y\,\partial x} + y^2\frac{\partial^2 U}{\partial y^2} + y.1.\frac{\partial U}{\partial y}$$

Hence

$$x^2\frac{\partial^2}{\partial x^2}U + 2xy\frac{\partial^2}{\partial x\partial y}U + y^2\frac{\partial^2}{\partial y^2}U + \left[x\frac{\partial U}{\partial x} + y\frac{\partial U}{\partial y}\right] = 4U$$

Since the terms in the square brackets equal $2U$ from (i), the result follows.

(c) $\dfrac{\mathrm{d}z}{\mathrm{d}t} = \dfrac{\partial z}{\partial x}\cdot\dfrac{\mathrm{d}x}{\mathrm{d}t} + \dfrac{\partial z}{\partial y}\cdot\dfrac{\mathrm{d}y}{\mathrm{d}t}$

$$= \left[\frac{3x^2 - (x^3 - 4y^2).2x}{\exp(x^2 + y^2)}\right]\frac{\mathrm{d}x}{\mathrm{d}t} + \left[\frac{-8y - (x^3 - 4y^2).2y}{\exp(x^2 + y^2)}\right]\frac{\mathrm{d}y}{\mathrm{d}t}$$

Since

$$y = x, \frac{\mathrm{d}y}{\mathrm{d}t} = \frac{\mathrm{d}x}{\mathrm{d}t} = 3.$$

(i) When $x = 1$ ($= y$) then

$$\frac{\mathrm{d}z}{\mathrm{d}t} = \frac{3}{e^2}\,[3 - (-3).2 - 8 - (-3).2]$$

$$= \underline{21e^{-2}}$$

(ii) When x is large then $x^3 - 4y^2 \simeq x^3$, therefore

$$\frac{\mathrm{d}z}{\mathrm{d}t} \simeq 3e^{-2x^2}[0 - x^3.2x - 0 - x^3.2x] \simeq \underline{0-}$$

Example 30

(a) The potential ϕ at a point (r, θ) due to a sphere of relative permittivity k and radius a in a uniform electric field E is defined by

$$\phi = \begin{cases} -Er\cos\theta + \dfrac{k-1}{k+2} \cdot \dfrac{Ea^3}{r^2}\cos\theta & \text{if } r \geqslant a \\[2ex] -\dfrac{3\,Er\cos\theta}{k+2} & \text{if } r < a \end{cases}$$

Show that ϕ and $\partial\phi/\partial\theta$ are continuous in r for all values of θ, and that $\partial\phi/\partial r$ is discontinuous in r at $r = a$.

(b) The magnitude of the field S is given by

$$S^2(r, \theta) = \left(\frac{\partial\phi}{\partial r}\right)^2 + \frac{1}{r^2}\left(\frac{\partial\phi}{\partial\theta}\right)^2$$

Show that
  (i) S is constant inside the sphere, $r < a$
  (ii) $S \to E$ as $r \to \infty$, and

  (iii) $S = \dfrac{3E}{k+2}\sqrt{k^2\cos^2\theta + \sin^2\theta}$ on the outer surface of the

  sphere. Hence find the maximum value of S, assuming that $k > 1$.   (CEI)

(a) The formulae for ϕ comprise continuous functions so that each part is continuous.
    At $r = a$,

$$\lim_{r\to a+}\phi = -Ea\cos\theta + \left(\frac{k-1}{k+2}\right)Ea\cos\theta = \left[\frac{-(k+2)+(k-1)}{(k+2)}\right]\cos\theta$$

$$\lim_{r\to a-}\phi = -\frac{3Ea\cos\theta}{(k+2)}$$

Hence ϕ is continuous for all values of r.
    The same conclusion for $\partial\phi/\partial\theta$ follows since, in effect, $\cos\theta$ is replaced by $(-\sin\theta)$. For $r > a$,

$$\frac{\partial\phi}{\partial r} = -E\cos\theta - \frac{2(k-1)}{(k-2)} \cdot \frac{Ea^3\cos\theta}{r^3}$$

$$\to -\frac{3Ek\cos\theta}{(k+2)} \quad \text{as} \quad r \to a+$$

For $r < a$,

$$\frac{\partial \phi}{\partial r} = -\frac{3E \cos \theta}{(k+2)} \rightarrow -\frac{3E \cos \theta}{(k+2)} \quad \text{as} \quad r \rightarrow a-$$

Hence $\partial \phi / \partial r$ is discontinuous at $r = a$.

(b)
$$S^2(r, \theta) = \begin{cases} E^2 \cos^2\theta \left[1 + \frac{2(k-1)}{(k+2)} \frac{a^3}{r^3} \right]^2 + E^2 \sin^2\theta \left[1 - \frac{(k-1)}{(k+2)} \frac{a^3}{r^3} \right]^2, \\ \qquad\qquad\qquad\qquad\qquad\qquad\qquad\qquad\qquad r \geqslant a \\ E^2 \cos^2\theta \cdot \frac{9}{(k+2)^2} + E^2 \sin^2\theta \cdot \frac{9}{(k+2)^2}, \qquad r < a \end{cases}$$

(i) Inside the sphere $S^2 = \dfrac{9E^2}{(k+2)^2}$, constant

(ii) As $r \rightarrow \infty$ $(> a)$ $S^2 \rightarrow E^2 \cos^2\theta + E^2 \sin^2\theta = E^2$

(iii) As $r \rightarrow a+$ $(> a)S^2 \rightarrow E^2 \cos^2\theta \left[\dfrac{3k}{k+2} \right]^2 + E^2 \sin^2\theta \left[\dfrac{3}{k+2} \right]^2$ \qquad (1)

$$\text{i.e. } S \rightarrow \left(\frac{3E}{k+2} \right) [k^2 \cos^2\theta + \sin^2\theta]^{1/2}$$

as required.

Differentiating (1) w.r.t. θ

$$2S \frac{\partial S}{\partial \theta} = 0 \quad \text{when} \quad \frac{9E^2}{(k+2)^2} \left[-2k^2 \cos \theta \sin \theta + 2 \sin \theta \cos \theta \right] = 0$$

This is when $\cos \theta = 0$ or $\sin \theta = 0$.
When $\cos \theta = 0$, $\sin^2\theta = 1$ and

$$S = \frac{3E}{(k+2)}.$$

When $\sin \theta = 0$, $\cos^2\theta = 1$ and

$$S = \frac{3kE}{(k+2)}.$$

The value as $r \rightarrow \infty$ is $S = E$ and the value inside the sphere is

$$S = \frac{3E}{(k+2)}.$$

Hence, since $k > 1$, the maximum value of S is $\underline{3kE/(k+2)}$

G

Integration

Example 31

A point body, P, of mass m is placed at a distance L along the normal from the centre 0 of a very thin circular disc of radius R. This disc, which has a uniform thickness T, has a density ρ per unit volume. It is required to find the gravitational attraction of the disc on the body. Refer to Figure 29.

It may be assumed that the force of attraction between two bodies of mass m_1 and m_2 a distance d apart is given by Gm_1m_2/d^2, where G is a constant. (CEI)

The mass of the element shown is $2\pi r\,\delta r\,T\rho$.

The force between the mass at P and the element is

$$\frac{G \cdot 2\pi r\,\delta r T\rho \cdot m}{d^2}$$

and when resolved along the axis OP has component

$$\frac{G \cdot 2\pi r\,\delta r\,T\rho m}{d^2} \cos\theta$$

where $\cos\theta = L/d$ and $d = \sqrt{L^2 + r^2}$.

Hence the component of force is

$$\frac{G.2\pi L\,T\rho\,mr}{(L^2 + r^2)^{3/2}}\,\delta r.$$

The total force due to all elements is found by summation and then,

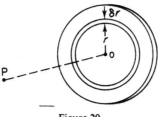

Figure 29

taking the limit as $\delta r \to 0$ we obtain

$$F = \int_0^R \frac{G . 2\pi T\rho \, mLr}{(L^2 + r^2)^{3/2}} \, dr.$$

Substituting $p = L^2 + r^2$ so that $dp/dx = 2r$, we note that

$$\int \frac{2r\,dr}{(L^2 + r^2)^{3/2}} = \int \frac{dp}{p^{3/2}} = [-2p^{-1/2}] = [-2(L^2 + r^2)^{-1/2}].$$

Hence the substitution $L^2 + r^2 = p$ eventually leads to

$$F = \left[-\frac{G . 2\pi T\rho \, mL}{\sqrt{L^2 + r^2}} \right]_0^R$$

$$= 2\pi T\rho \, mG \left(1 - \frac{L}{\sqrt{L^2 + R^2}} \right).$$

Example 32

(a) Find the length of one arch of the cycloid

$$x = a(\theta - \sin \theta), \qquad y = a(1 - \cos \theta).$$

(b) Find the coordinates of the centroid of the area enclosed by the curve $y = 8/(x^2 + 4)$, the x-axis and the ordinates $x = -2\sqrt{3}$ and $x = 2\sqrt{3}$.

(c) Find the second moment of area about both axes of the area between the curves $y = x^{1/2}$ and $y = x$. (LUT)

(a) Since the equation of the cycloid is given in terms of the parameter θ, it makes sense to evaluate the length as an integral over θ. See Figure 30(a).

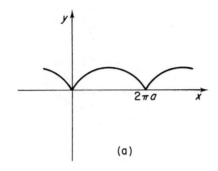

(a)

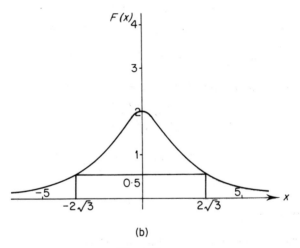

(b)

Figure 30

From the periodicity of the curve it seems reasonable to suppose that the length we want is that of the arch between $x = 0$ and $x = 2a\pi$. The formula for the length of arc is

$$L = \int_0^{2a\pi} \sqrt{1 + \left(\frac{dy}{dx}\right)^2} \, dx.$$

Converting this in terms of θ we have

$$L = \int_0^{2\pi} \sqrt{\left(\frac{dx}{d\theta}\right)^2 + \left(\frac{dy}{\delta\theta}\right)^2} \, . \, d\theta$$

$$= \int_0^{2\pi} \sqrt{a^2(1 - \cos \theta)^2 + a^2 \sin^2\theta} \; d\theta$$

$$= a \int_0^{2\pi} \sqrt{2 - 2 \cos \theta} \; d\theta$$

$$= a\sqrt{2} \int_0^{2\pi} \sqrt{2 \sin^2\theta/2} \; d\theta = \underline{8a}$$

(b) Refer to Figure 30(b).

By symmetry, the area is given by

$$A = 2 \int_0^{2\sqrt{3}} \frac{8}{x^2 + 4} \; dx$$

$$= 16 \left[\frac{1}{2} \tan^{-1} \frac{x}{2} \right]_0^{2\sqrt{3}}$$

$$= \frac{8\pi}{3}.$$

By symmetry again, the centroid lies on the y-axis; hence $\bar{x} = 0$.

We take the moment of area about the x-axis of that part of the curve enclosed by the x-axis and the given ordinates. We divide the area into two parts: the rectangle from the x-axis to the line $y = \frac{1}{2}$ (A_1, say) and the remainder (A_2).

For A_1: Length of a strip $= 4\sqrt{3}$, thickness $= dy$.

Hence, moment $= \displaystyle\int_0^{1/2} y \cdot 4\sqrt{3} \cdot dy = \frac{\sqrt{3}}{2}.$

For A_2: Length of a strip $= 2x = 2\left(\dfrac{8}{y} - 4\right)^{1/2}.$

Hence, moment $= \displaystyle\int_{1/2}^2 y \cdot 2\left(\frac{8}{y} - 4\right)^{1/2} dy$

$$= 4 \int_{1/2}^2 (2y - y^2)^{1/2} \, dy$$

If we write $2y - y^2 = 1 - (1 - y)^2$ and substitute $1 - y = \sin \theta$ we need to be careful for two reasons. When choosing suitable limits

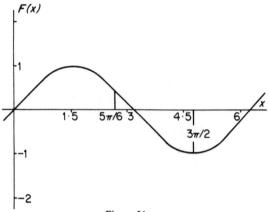

Figure 31

on θ, we want a continuous substitution; the range $[5\pi/6, 3\pi/2]$ for θ gives a continuous change in y from $1/2$ to 2; see Figure 31. However, using

$$\mathrm{d}y/\mathrm{d}\theta = -\cos\theta$$

leads to the result that moment $= 4\int_{5\pi/6}^{3\pi/2}\cos\theta . (-\cos\theta\,\mathrm{d}\theta)$ which in turn leads to a negative value for the moment. The catch here is that the integrand is a positive square root whereas $\cos\theta$ is negative on the interval $[5\pi/6, 3\pi/2]$. Hence we should write the moment as

$$4\int_{5\pi/6}^{3\pi/2}(-\cos\theta) . (-\cos\theta\,\mathrm{d}\theta) = 4\pi/3 + \sqrt{3}/2$$

Therefore the total moment $= (4\pi/3) + \sqrt{3} = $ (total area) $\times \bar{y}$ which leads to

$$\bar{y} = \frac{1}{2} + \frac{3\sqrt{3}}{8\pi}$$

It is always worth checking that the result is reasonable for the geometry of the area concerned.

(c)   Refer to Figure 32.

The area enclosed by the curves is

$$A = \int_0^1 (x^{1/2} - x)\,\mathrm{d}x = \frac{1}{6}$$

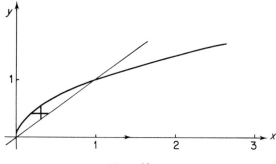

Figure 32

To find the second moment of area about the y-axis we take strips parallel to that axis. The second moment of area of such a strip is $(x^{1/2} - x)\, \mathrm{d}x \cdot x^2$ and therefore the total second moment is

$$\int_0^1 x^2(x^{1/2} - x)\, \mathrm{d}x = \underline{\frac{1}{28}}$$

To find the second moment of area about the x-axis we take strips parallel to it. For such a strip the s.m.a. is $(y - y^2)\cdot \mathrm{d}x \cdot y^2$ and therefore the total second moment is

$$\int_0^1 y^2(y - y^2)\, \mathrm{d}y = \underline{\frac{1}{20}}.$$

Example 33

(a) In an electrostatic potential problem,

$$V(x) = \int_0^\pi \frac{a^2 \sigma \sin \phi}{2\varepsilon(a^2 + x^2 - 2ax \cos \phi)^{1/2}}\, \mathrm{d}\phi$$

where a, σ, ε are constants and $x > a$ is independent of ϕ.

Evaluate this integral and thus obtain a simple formula for $V(x)$.

(b) Find the area under the cycloid given by

$$x = a(\theta - \sin \theta), \qquad y = a(1 - \cos \theta)$$

between the points $\theta = 0$ and $\theta = \pi$.　(CEI)

(a) Note that x is a fixed value and that ϕ is the variable of integration.

To simplify matters, put

$$A = \frac{a^2 \sigma}{2\varepsilon}, \qquad B = a^2 + x^2, \qquad C = 2ax$$

Then

$$V(x) = \int_0^\pi \frac{A \sin \phi}{(B - C \cos \phi)^{1/2}} \, d\phi = \frac{2A}{C} \left[(B - C \cos \phi)^{1/2} \right]_0^\pi$$

$$= \frac{2A}{C} \left[(B + C)^{1/2} - (B - C)^{1/2} \right]$$

$$= \frac{2A}{C} \left[(a^2 + x^2 + 2ax)^{1/2} - (a^2 + x^2 - 2ax)^{1/2} \right]$$

$$= \frac{a\sigma}{2\varepsilon x} \left[(a + x) - (x - a) \right] = \frac{a^2 \sigma}{\varepsilon x}$$

(Note that $x > a \Rightarrow x - a > 0$ and $(\ )^{1/2}$ implies a positive value.)

(b) From Figure 33,

$$\text{Area} = \int_{\theta=0}^{\theta=\pi} y \cdot \frac{dx}{d\theta} \cdot d\theta = \int_0^{\pi a} y \, dx \quad \text{and} \quad \frac{dx}{d\theta} = a(1 - \cos \theta)$$

Therefore,

$$\text{Area} = \int_0^\pi a^2 (1 - \cos \theta)^2 \, d\theta$$

$$= a^2 \int_0^\pi \left[1 - 2 \cos \theta + \tfrac{1}{2}(1 + \cos 2\theta) \right] \, d\theta$$

$$= \frac{3}{2} \pi a^2.$$

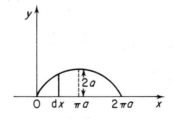

Figure 33

It is straightforward to see that since $\cos \theta$ integrates to $\sin \theta$ and $\cos 2\theta$ to $\frac{1}{2} \sin 2\theta$, which vanish at 0 and π, neither will contribute to the value of the definite integral.

Example 34

The shaded area in Figure 34 is defined by the intersection of the ellipse

$$\frac{x^2}{a^2} + \frac{y^2}{b^2} = 1$$

and the rectangular hyperbola $xy = 1$.

Prove that the abscissae x_1 and x_2 are real if $ab > 2$, and derive expressions for x_1 and x_2. Hence find in terms of x_1 and x_2 the area between the curves and the volume of revolution of this area about OX.

Evaluate the volume numerically given that $a = \sqrt{5}$ and $b = 1$.

(CEI)

The intersection occurs where

$$\frac{x^2}{a^2} + \frac{1}{b^2 x^2} = 1$$

or

$$b^2 (x^2)^2 - a^2 b^2 x^2 + a^2 = 0$$

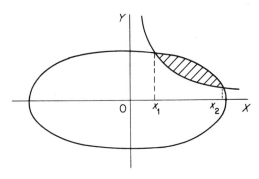

Figure 34

Hence

$$x^2 = \frac{a^2b^2 \pm \sqrt{a^4b^4 - 4a^2b^2}}{2b^2}$$

For real x^2 then we require $a^4b^4 - 4a^2b^2 \geqslant 0$, i.e. $ab \geqslant 2$.

In this case the intersection points are

$$x_1 = \left(\frac{a^2}{2} - \frac{a}{2b}\sqrt{a^2b^2 - 4}\right)^{1/2} \quad \text{and} \quad x_2 = \left(\frac{a^2}{2} + \frac{a}{2b}\sqrt{a^2b^2 - 4}\right)^{1/2}$$

The area shaded is

$$A = \int_{x_1}^{x_2} \{y_{\text{ellipse}} - y_{\text{hyperbola}}\}\, \mathrm{d}x$$

$$= \int_{x_1}^{x_2} \left\{b\sqrt{1 - \frac{x^2}{a^2}} - \frac{1}{x}\right\}\, \mathrm{d}x$$

Put $x = a\sin\theta$ so that

$$A = \int_{x=x_1}^{x=x_2} \{b\cos\theta\}\, a\cos\theta\, \mathrm{d}\theta - [\log_e x]_{x_1}^{x_2}$$

$$= ab \int_{x=x_1}^{x=x_1} \tfrac{1}{2}(1 + \cos 2\theta)\, \mathrm{d}\theta - [\log_e x]_{x_1}^{x_2}$$

$$= ab[\tfrac{1}{2}\theta + \tfrac{1}{4}\sin 2\theta]_{x=x_1}^{x=x_2} - [\log_e x]_{x_1}^{x_2}$$

$$= \underline{\left[\frac{ab}{2}\sin^{-1}\left(\frac{x}{a}\right) + \frac{ab}{2} \cdot \frac{x}{a}\sqrt{1 - \frac{x^2}{a^2}} - \log_e x\right]_{x_1}^{x_2}}$$

$$\text{Volume} = \int_{x_1}^{x_2} \pi(y_{\text{ellipse}}^2 - y_{\text{hyperbola}}^2)\, \mathrm{d}x$$

$$= \pi \int_{x_1}^{x_2}\left[b^2\left(1 - \frac{x^2}{a^2}\right) - \frac{1}{x^2}\right]\, \mathrm{d}x = \pi\left[b^2 x - \frac{x^3}{3a^2} + \frac{1}{x}\right]_{x_1}^{x_2}$$

$$= \underline{\pi b^2\left\{x_2\left(1 - \frac{x_2^2}{3a^2}\right) - x_1\left(1 - \frac{x_1^2}{3a^2}\right)\right\} + \pi\left(\frac{1}{x_2} - \frac{1}{x_1}\right)}$$

When

$$a = \sqrt{5}, \ b = 1; \ x_2, x_1 = \left(\frac{5}{2} \pm \frac{\sqrt{5}}{2}\right)^{1/2} = 1.902, 1.176.$$

Hence

$$\text{Volume} \simeq 0.168\pi \simeq \underline{0.528}.$$

Example 35

(a) The profile of an impeller blade is bounded by the lines $x = 0.1$, $y = 2x$, $y = e^{-x}$, $x = 1$ and the x-axis. The blade thickness t varies linearly with x, thus $t = (1.1 - x)T$, where T is a constant.

Find, to two decimal places from tables, or otherwise, the point given by the equation $2x = e^{-x}$.

Hence determine the volume of the blade, showing the result to be approximately $\frac{1}{4} T$.

(b) A wave is defined by the equation

$$e = E_1 \sin \omega t + E_3 \sin 3\omega t$$

where E_1, E_3, and ω are constants.

Find the root mean square (r.m.s.) value of e over the interval $0 \leqslant t \leqslant (\pi/\omega)$. (CEI)

(a) Figure 35 shows the blade profile. The coordinates of A are found by first solving $2x = e^{-x}$ to obtain $x = 0.35$ (2dp) and hence $A = (0.35, 0.70)$. (The iteration formula $x_{n+1} = 0.5 \, e^{-x_n}$ readily yields the required value of 0.35.)

Then the volume of the blade

$$V = \int_{0.1}^{1} y.t.\mathrm{d}x$$

$$= \int_{0.1}^{0.35} 2x(1.1 - x)T \, \mathrm{d}x + \int_{0.35}^{1} e^{-x}(1.1 - x)T \, \mathrm{d}x$$

$$= 0.25076T$$

$$\simeq \underline{0.25T}$$

(bear in mind that 0.35 is an approximate value).

(b) $e_{\text{rms}}^2 = \dfrac{\omega}{\pi} \displaystyle\int_0^{\pi/\omega} e^2 \, \mathrm{d}t$, by definition of root mean square.

$$= \frac{\omega}{\pi} \int_0^{\pi/\omega} \{E_1^2 \sin^2 \omega t + E_3^2 \sin^2 3\omega t + 2E_1 E_3 \sin \omega t \sin 3\omega t\} \, \mathrm{d}t$$

But

$$\int_0^{\pi/\omega} \sin^2 \omega t \, \mathrm{d}t = \frac{1}{2} \int_0^{\pi/\omega} (1 - \cos 2\omega t) \, \mathrm{d}t = \frac{1}{2} \frac{\pi}{\omega} - 0$$

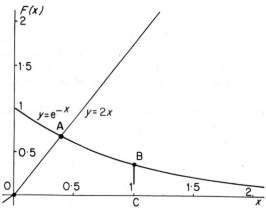

Figure 35

and

$$\int_0^{\pi/\omega} \sin^2 3\omega t \, dt = \frac{1}{2}\frac{\pi}{\omega}$$

and

$$\int_0^{\pi/\omega} 2 \sin \omega t \sin 3\omega t \, dt = \int_0^{\pi/\omega} \{\cos 2\omega t - \cos 4\omega t\} \, dt = 0$$

Therefore

$$e^2_{\text{rms}} = \frac{\omega}{\pi}\left[\frac{E_1^2 + E_3^2}{2}\right] \cdot \frac{\pi}{\omega}$$

so that

$$e_{\text{rms}} = \frac{1}{\sqrt{2}}(E_1^2 + E_3^2)^{1/2}$$

Example 36

A cardioid is defined by the polar equation

$$r = a(1 + \cos \theta)$$

Given that the area of the triangle OPQ in Figure 36 is $\frac{1}{2}r^2 \, d\theta$, find the area enclosed by the cardioid.

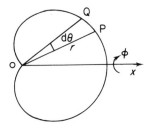

Figure 36

If the triangle OPQ is rotated about Ox by a small angle dϕ show that the volume of the element so generated is $\frac{1}{3}r^3 \sin \theta\, d\theta\, d\phi$. Deduce the total volume generated by OPQ when rotated about Ox through an angle 2π, and hence show that the volume of the cardioid of revolution is $\frac{8}{3}\pi a^3$. (CEI)

Using the symmetry of the cardioid about the x-axis, its area is given by

$$A = \int_0^{2\pi} \frac{1}{2} r^2\, d\theta = 2 \int_0^{\pi} \frac{1}{2} a^2 (1 + \cos \theta)^2\, d\theta$$

$$= \underline{\frac{3\pi}{2} a^2}.$$

The volume of the rotated element (assumed conical)

$$= \frac{1}{3}(\text{base area}) \times (\text{height})$$
$$= \frac{1}{3}(r\, d\theta . r \sin \theta\, d\phi) . r$$
$$= \underline{\frac{1}{3} r^3 \sin \theta\, d\theta\, d\phi}$$

Hence the volume of the conical shell obtained by integrating over ϕ is

$$\int_0^{2\pi} (\tfrac{1}{3} r^3 \sin \theta\, d\theta)\, d\phi = \tfrac{2}{3}\pi r^3 \sin \theta\, d\theta.$$

Finally, the volume of revolution of the cardioid is

$$\int_0^{\pi} \tfrac{2}{3}\pi r^3 \sin \theta\, d\theta = \underline{\tfrac{8}{3}\pi a^3}.$$

The limits on this integral are 0 and π since we rotate that part of the cardioid above the x-axis only.

H

Differential Equations

Example 37

(a) A mass m, constrained to move along the x-axis, is attracted towards the origin with a force proportional to its distance from the origin. Find the motion

   (i) if it starts from rest at $x = x_0$, and

   (ii) if it starts from the origin with velocity v_0.

(b) Solve the differential equation

$$\frac{d^4 y}{dx^4} = 16y$$

(CEI)

(c) If $(dy/dx) + 2y \tan x = \sin x$

and $y = 0$ when $x = \pi/3$, show that the maximum value of y is $\frac{1}{8}$.

(CEI)

(a) The equation of motion is

$$m\ddot{x} = -mk^2 x$$

where k is constant. This has general solution

$$x = A \cos kt + B \sin kt; \qquad \dot{x} = -kA \sin kt + kB \cos kt$$

(i) at $t = 0$, $\dot{x} = 0$, $x = x_0$, so that $B = 0$, $A = x_0$ and $\underline{x = x_0 \cos kt}$

(ii) at $t = 0$, $x = 0$, $\dot{x} = v_0$ so that $A = 0$, $kB = v_0$ and $\underline{x = (v_0/k)\sin kt}$

(b) The equation $(d^4 y/dx^4) - 16y = 0$ can be solved in a similar way to a second order differential equation.

Substitution of $y = e^{mx}$ produces the auxiliary equation

$$m^4 - 16 = 0$$

which has roots $m = 2, -2, 2i, -2i$ and basic solutions $y = e^{2x}, e^{-2x}, e^{2ix}, e^{-2ix}$. The last two can be combined to give a mixture of $\cos 2x$ and $\sin 2x$ since

$$E\,e^{2ix} + F\,e^{-2ix} \equiv E \cos 2x + Ei \sin 2x$$
$$+ F \cos 2x + Fi \sin 2x$$

and the general solution to the given equation is

$$\underline{y = A\,e^{2x} + B\,e^{-2x} + C \cos 2x + D \sin 2x}$$

where A, B, C, and D are constants.

(c) This is an equation which can be solved via the integrating factor method.

$$\log[\text{I.F.}] = \int 2 \tan x \, dx = -2 \log \cos x = \log(\sec^2 x)$$

therefore

$$\text{I.F.} = \sec^2 x$$

Multiplying the differential equation term by term by $\sec^2 x$ and rearranging the left-hand side produces

$$\frac{d}{dt}(y \sec^2 x) = \sin x \sec^2 x$$

therefore

$$y \sec^2 x = \int \sin x \sec^2 x \, dx = \int \sec x \tan x$$
$$= \sec x + C, \qquad C \text{ constant}$$

therefore

$$y = \cos x + C \cos^2 x$$

Use of the initial condition shows that $C = -2$ and therefore

$$y = \cos x - 2 \cos^2 x$$

so that

$$y' = -\sin x + 4 \sin x \cos x$$

and

$$y'' = -\cos x + 4(\cos^2 - \sin^2 x)$$

When

$$y' = 0, \ \sin x(4 \cos x - 1) = 0$$

When

$$\sin x = 0, \qquad \cos x = \pm 1;$$
$$\cos x = 1 \Rightarrow y'' = 3, \qquad \cos x = -1 \Rightarrow y'' = 5.$$

When

$$\cos x = \tfrac{1}{4}, \qquad \sin x = \pm \sqrt{15}/4;$$
$$\sin x = \pm \sqrt{15}/4 \Rightarrow y'' = -29/8.$$

Hence there are local maxima where $\cos x = \tfrac{1}{4}$ and these have value

$$y_{max} = \tfrac{1}{4} - 2 \cdot \tfrac{1}{16} = \tfrac{1}{8}.$$

Example 38

A reservoir has been contaminated by effluent from a factory. The capacity of the reservoir is 10^6 litres. The degree of contamination is 0.02% by weight. The (constant) average daily rate of consumption of water for non-drinking purposes is 2×10^4 litres and this is continuously replaced by pure water. How long will it be before the concentration of contaminant drops to the safe level of 10^{-5}%?

(CEI)

We first make the assumption that the contaminant is at all times uniformly dispersed through the reservoir; how else can we tackle this problem? Then we assume that the changes that take place during one day can be adequately described by the rule:

$$\text{amount of contaminant removed} = \frac{2 \times 10^4}{10^6}$$

$$\times \text{(amount at start of the day)}$$

It is easier in practice to deal with mass, m; this is proportional to the concentration c since the volume of the reservoir is assumed to be fixed. Let m_n be the mass of contaminant in the reservoir after n days. Then

$$m_1 = \left(\frac{49}{50}\right) m_0, \quad m_2 = \left(\frac{49}{50}\right) m_1 = \left(\frac{49}{50}\right)^2 m_0$$

and, in general

$$m_n = \left(\frac{49}{50}\right)^n m_0.$$

Hence the time taken for the concentration to drop to the required level is n days where n is given by

$$\frac{c_0}{c_n} = \frac{m_0}{m_n} = \left(\frac{50}{49}\right)^n = \frac{0.02}{10^{-5}} = 2000$$

Therefore

$$n = 376.$$

Were we to try a differential equation model with $\delta t = 1$ day (permissible so long as $\delta t \ll$ the time for the concentration to drop to the required level), then

$$\frac{\mathrm{d}m}{\mathrm{d}t} = -\frac{m}{50}$$

(Note the negative sign; m decreases as t increases.)
Therefore

$$\frac{1}{m}\frac{\mathrm{d}m}{\mathrm{d}t} = -\frac{1}{50} \quad \text{and} \quad \int \frac{1}{m}\frac{\mathrm{d}m}{\mathrm{d}t}\,\mathrm{d}t = \int -\frac{1}{50}\,\mathrm{d}t$$

so that

$$\log_e m = -\frac{t}{50} + A, \qquad A \text{ constant.}$$

It follows that

$$m = m_0\, e^{-t/50}$$

and the required time is t days where

$$t = 50 \log_e 2000 = \underline{380}$$

The agreement between the two approaches is quite good.

Example 39

A room of volume V m^3 is supplied with fresh air containing some CO_2. Let $c(t)$ be the concentration of CO_2 in the room at time $t > 0$, c_0 the initial concentration and c_f the concentration in the fresh air supply; all three are measured in parts per 10 000. Let Q m^3/sec be the rate of fresh air supplied and Q_p m^3/sec the volume of CO_2 produced by people inside the room.

Derive a differential equation for the concentration $c(t)$ and solve it.

A room of volume 170 m^3 receives a total change of air every 30 minutes and has a CO_2 content of 0.03% without people present. The concentration of CO_2 in the outside air is also 0.03%. If the production of CO_2 per person is 4.7×10^{-6} m^3/sec, what is the maximum number of people allowed in the room at any one time if the concentration of CO_2 in the room is not to exceed 0.1%? Sketch the solution curve in this case for $c(t)$. (Hint: the long-term solution only is important.)

(LUT)

(a) Where concentration is the dependent variable it is usually best to work in terms of mass or volume. Here we deal with volume of CO_2. The fraction of CO_2 in the incoming air is $c_f/10^4$.

$$\text{Rate of input: } \left(\frac{Qc_f}{10^4} + Q_p\right) \text{m}^3/\text{sec}$$

$$\text{Rate of output: } \frac{Qc}{10^4} \text{m}^3/\text{sec}$$

therefore

$$\text{Rate of change of mass} = \left(\frac{Qc_f + 10^4 Q_p - Qc}{10^4}\right) \text{m}^3/\text{sec}$$

But mass of CO_2 at any time $= (c/10^4) V$ therefore

$$\text{Rate of change of mass} = \frac{\text{d}}{\text{d}t}\left(\frac{c}{10^4} V\right) = \frac{V}{10^4} \frac{\text{d}c}{\text{d}t}$$

Equating the two expressions and rearranging, we have

$$\frac{\text{d}c}{\text{d}t} + \frac{Q}{V} c = \frac{Qc_f + 10^4 Q_p}{V} = Q^*, \text{ say}$$

(Note that Q^* is constant.)

The integrating factor of $e^{+Qt/V}$ produces a solution

$$c = Q^* \frac{V}{Q} + \text{constant} (e^{-Qt/V})$$

and, on applying the initial condition, we obtain the formula

$$c = c_0\, e^{-Qt/V} + (Qc_f + 10^4\, Q_p)(1 - e^{-Qt/V})/Q.$$

(b) Let the number of people in the room be N. Then their rate of production of CO_2 is

$$Q_p = 4.76 \times 10^{-6} N \qquad m^3/sec.$$

The long-term solution is

$$c = \frac{(Qc_f + 10^4 Q_p)}{Q} = c_f + \frac{10^4 Q_p}{Q}$$

$$= 3 + \frac{10^4 \cdot 4.76 \times 10^{-6}\, N}{170/(30 \times 60)} < 10$$

(Remember that concentration is to be quoted in parts per 10^4.)
Therefore

$$N < \frac{7 \times 170}{1800 \times 10^4 \times 4.76 \times 10^{-6}} \cong 13.9$$

We could probably get by with 14 people in the room.

Example 40

An object moves with constant velocity V along DT (parallel to the y-axis) and, as it passes D, a missile is launched from 0. The missile travels with constant speed nV, where $n > 1$, and is continuously directed towards the target object. $P(x, y)$ and $T(d, s/n)$ are the current positions of missile and target respectively, and $s = $ arc OP. Refer to Figure 37. Show that the path of P is given by the differential equation

$$n(d - x)\frac{d^2y}{dx^2} = \sqrt{1 + \left[\frac{dy}{dx}\right]^2}$$

By substituting $p = dy/dx$, integrate and solve for p and hence obtain $y(x)$.

Deduce that the missile reaches the object after a time $nd/(n^2 - 1)V$.

What is the minimum missile speed which would guarantee a hit within a distance R from the origin 0?

(CEI)

See Figure 38.

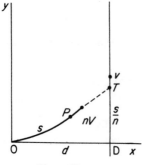

Figure 37   Figure 38

From $\triangle$PTE,

$$\frac{dy}{dx} = \frac{TE}{PE} = \frac{s/n - y}{d - x} \qquad (1)$$

Differentiate (1) w.r.t. x to obtain

$$(d - x)\frac{d^2y}{dx^2} - \frac{dy}{dx} = \frac{1}{n}\frac{ds}{dx} - \frac{dy}{dx}$$

But

$$\left(\frac{ds}{dx}\right)^2 = 1 + \left(\frac{dy}{dx}\right)^2,$$

so that

$$n(d - x)\frac{d^2y}{dx^2} = \sqrt{1 + \left(\frac{dy}{dx}\right)^2}$$

Substitute $p = dy/dx$, note that $(dp/dx) = d^2y/dx^2$ and separate the variables to obtain

$$\int \frac{dp}{\sqrt{1 + p^2}} = \int \frac{dx}{n(d - x)}.$$

integrating,

$$\log_e(p + \sqrt{1 + p^2}) = -\frac{1}{n}\log(d - x) + \log A, \qquad A \text{ constant.}$$

or

$$(p + \sqrt{1 + p^2})(d - x)^{1/n} = A.$$

At $x = 0$, $p = 0$ so that $A = d^{1/n}$. Hence

$$\frac{dy}{dx} = p = \frac{1}{2}\left[\left(\frac{d}{d-x}\right)^{1/n} - \left(\frac{d}{d-x}\right)^{-1/n}\right]$$

$$= \frac{1}{2}\left[-\left(\frac{d-x}{d}\right)^{1/n} + \left(\frac{d-x}{d}\right)^{-1/n}\right]$$

{This follows from the argument that if

$$p + \sqrt{1+p^2} = \alpha$$

then

$$\sqrt{1+p^2} = \alpha - p$$

and

$$1 + p^2 = \alpha^2 - 2\alpha p + p^2$$

hence

$$p = (\alpha^2 - 1)/2\alpha = \tfrac{1}{2}(\alpha - \alpha^{-1})\}$$

Integrating, and using the condition $x = 0$, $y = 0$,

$$y = B + \left[\frac{1}{2}\left(\frac{d}{1+\frac{1}{n}}\right)\left(\frac{d-x}{d}\right)^{1+1/n} - \left(\frac{d}{1-\frac{1}{n}}\right)\left(\frac{d-x}{d}\right)^{1-1/n}\right]$$

where

$$B = \frac{nd}{n^2 - 1}.$$

The missile reaches the object when $x = d$ (and, therefore, $y = B$) after a time

$$\frac{B}{V} = \frac{nd}{(n^2 - 1)V}.$$

To gaurantee a hit within a distance R from 0,

$$B \leqslant \sqrt{R^2 - d^2}$$

i.e.

$$n^2 - 1 \geqslant \frac{nd}{\sqrt{R^2 - d^2}}$$

i.e.

$$\left(n - \frac{d}{2\sqrt{R^2 - d^2}}\right)^2 \geqslant 1 + \frac{d^2}{4(R^2 - d^2)}$$

Therefore,

$$nV \geqslant V\left[\frac{d}{2\sqrt{R^2 - d^2}} + \left(\frac{4R^2 - 3d^2}{4(R^2 - d^2)}\right)^{1/2}\right]$$

Example 41

If y is a function of x and if $p = dy/dx$, show that

$$p \frac{dp}{dy} = \frac{d^2 y}{dx^2}$$

Hence show that the differential equation

$$EI \frac{d^2 y}{dx^2} + Fy \left[1 + \left(\frac{dy}{dx} \right)^2 \right]^{3/2} = 0$$

which relates to a beam with Young's modulus E, second moment of area of cross-section I and axial load F, can be reduced to a first order equation having the solution

$$p = \sqrt{ \left(C + \frac{Fy^2}{2EI} \right)^{-2} - 1 }$$

where C is an arbitrary constant.

If the maximum deflection d of the beam occurs when $dy/dx = 0$, show that

$$C = 1 - \frac{Fd^2}{2EI} \tag{CEI}$$

If $p = dy/dx$, differentiate w.r.t. y to obtain

$$\frac{dp}{dy} = \frac{d^2 y}{dx^2} \cdot \frac{dx}{dy} = \frac{d^2 y}{dx^2} \cdot \frac{1}{p}. \tag{1}$$

If we denote differentiation w.r.t. x by dashes, then

$$EI \, y'' + Fy \left[1 + (y')^2 \right]^{3/2} = 0. \tag{2}$$

Using the first result (1), this becomes

$$EI \, p \frac{dp}{dy} + Fy(1 + p^2)^{3/2} = 0.$$

Separating the variables leads to

$$EI \int \frac{p \, dp}{(1 + p^2)^{3/2}} = - F \int y \, dy$$

Hence

$$-EI(1 + p^2)^{-1/2} = -\tfrac{1}{2} Fy^2 - D, \qquad D \text{ constant}.$$

Then

$$(1 + p^2) = \left(\frac{Fy^2}{2EI} + C\right)^{-2}; \quad C = \frac{D}{EI}$$

Finally

$$p = \sqrt{\left(C + \frac{Fy^2}{2EI}\right)^{-2} - 1}$$

Since

$$y = d \quad \text{when } dy/dx = p = 0,$$

$$\left(C + \frac{Fy^2}{2EI}\right) = 1$$

hence the result.

Example 42

(a) Solve the differential equation

$$\frac{d^2y}{dx^2} - 2p\frac{dy}{dx} + y = 0$$

when (i) $p = \cos\theta$ and (ii) $p = \cosh\theta$, where θ is a constant in both cases. What is the most general solution when $\theta = 0$?

(b) Liquid is contained in a spherical vessel of radius a to a depth H. Find a formula for the volume of the liquid. At time $t = 0$, the liquid is allowed to leave the vessel via an orifice at the lowest point of the sphere at a rate proportional to the depth of the liquid in the vessel. Obtain an expression for the time taken to empty the vessel.   (CEI)

(a) Try a solution in the form $y = e^{mx}$.
Cancelling the common factor e^{mx}, the differential equation reduces to

$$m^2 - 2pm + 1 = 0.$$

hence

$$m = p \pm \sqrt{p^2 - 1}$$

(i) $p = \cos \theta \Rightarrow m = \cos \theta \pm i \sin \theta = e^{\pm i\theta}$.
Therefore

$$y = A \exp\{x\,e^{i\theta}\} + B \exp\{x\,e^{-i\theta}\}$$

where A, B are constants,
i.e.

$$\underline{y = e^{x\cos\theta}\{C \cos(x \sin \theta) + D \sin(x \sin \theta)\}}$$

where C, D are constants.

(ii) $p = \cosh \theta \Rightarrow m = \cosh \theta \pm \sinh \theta = e^{\pm \theta}$
Therefore

$$y = A \exp\{x\,e^{\theta}\} + B \exp\{x\,e^{-\theta}\}$$

i.e.

$$\underline{y = e^{x\cosh\theta}\{C \cosh(x \sinh \theta) + D \sin(x \sinh \theta)\}}$$

(Note that these formulae merely provide a means of calculating the value of y given a value of x.)
When $\theta = 0$, $m = 1$ as the only root.
Therefore

$$\underline{y = (Ax + B)\,e^{x}}$$

(b) The volume of liquid when it is at depth H is given by

$$V = \int_{a-H}^{a} \pi(a^2 - x^2)\,\mathrm{d}x = \pi H^2\left(a - \frac{H}{3}\right)$$

At some time $t > 0$ let the depth be h; then

$$\frac{\mathrm{d}V}{\mathrm{d}t} = -kh.$$

Also,

$$\frac{\mathrm{d}V}{\mathrm{d}t} = \frac{\mathrm{d}V}{\mathrm{d}h} \cdot \frac{\mathrm{d}h}{\mathrm{d}t}$$

so that

$$\left[2\pi h\left(a - \frac{h}{3}\right) - \pi\frac{h^2}{3}\right]\frac{\mathrm{d}h}{\mathrm{d}t} = -kh$$

(Note that we have first expressed V as a function of h.) Hence

$$(2\pi a - \pi h)\frac{dh}{dt} = -k$$

or

$$\frac{dt}{dh} = \frac{\pi(h-2a)}{k}$$

so that the time taken to empty the vessel is given by

$$t = \frac{\pi}{k}\int_H^0 (h-2a)\,dh = \frac{\pi}{k}H(2a - \tfrac{1}{2}H)$$

Example 43

In a radioactive series consisting of four different nuclides, starting with the parent substance N_1 and ending with the stable end product N_4, the amounts present at time t are given by

$$\frac{dN_1}{dt} = -\lambda_1 N_1 \qquad \frac{dN_2}{dt} = \lambda_1 N_1 - \lambda_2 N_2$$

$$\frac{dN_3}{dt} = \lambda_2 N_2 - \lambda_3 N_3 \qquad \frac{dN_4}{dt} = \lambda_3 N_3$$

λ_1, λ_2, λ_3 are different decay constants.

By means of Laplace transform method find $N_4(t)$, assuming that $N_1 = N_0$ (a constant) and $N_2 = N_3 = N_4 = 0$ at $t = 0$.　　　　(CEI)

Let the Laplace transforms of N_1, N_2, N_3, N_4 be $\bar{N}_1$, $\bar{N}_2$, $\bar{N}_3$, $\bar{N}_4$ respectively. Then transforming the equations produces in turn

(i)　$s\bar{N}_1 - N_0 = -\lambda_1\bar{N}_1$,　　　　(ii)　$s\bar{N}_2 = \lambda_1\bar{N}_1 - \lambda_2\bar{N}_2$

(iii)　$s\bar{N}_3 = \lambda_2\bar{N}_2 - \lambda_3\bar{N}_3$,　　　(iv)　$s\bar{N}_4 = \lambda_3\bar{N}_3$

i.e.

$$\bar{N}_1 = \frac{N_0}{s+\lambda_1}\left(\Rightarrow N_1 = N_0\,e^{-\lambda_1 t}\right)$$

$$\bar{N}_2 = \frac{\lambda_1\bar{N}_1}{s+\lambda_2} = \frac{\lambda_1 N_0}{(s+\lambda_2)(s+\lambda_1)}$$

$$\bar{N}_3 = \frac{\lambda_2\lambda_1 N_0}{(s+\lambda_3)(s+\lambda_2)(s+\lambda_1)}$$

and

$$\bar{N}_4 = \frac{\lambda_3\lambda_2\lambda_1 N_0}{s(s+\lambda_3)(s+\lambda_2)(s+\lambda_1)} \equiv \frac{A}{s} + \frac{B}{s+\lambda_3} + \frac{C}{s+\lambda_2} + \frac{D}{s+\lambda_1}$$

with A, B, C, D constants.

Via partial fractions we find that

$$\bar{N}_4 = \frac{N_0}{s} + \frac{N_0\lambda_1\lambda_2}{(s+\lambda_3)(\lambda_2-\lambda_3)(\lambda_3-\lambda_1)}$$

$$+ \frac{N_0\lambda_3\lambda_1}{(s+\lambda_2)(\lambda_1-\lambda_2)(\lambda_2-\lambda_3)} + \frac{N_0\lambda_2\lambda_3}{(s+\lambda_1)(\lambda_1-\lambda_2)(\lambda_3-\lambda_1)}$$

and hence

$$N_4 = N_0 + \frac{N_0}{(\lambda_1-\lambda_2)(\lambda_2-\lambda_3)(\lambda_3-\lambda_1)}$$

$$[\lambda_1\lambda_2(\lambda_1-\lambda_2)e^{-\lambda_3 t} + \lambda_3\lambda_1(\lambda_3-\lambda_1)e^{-\lambda_2 t} + \lambda_2\lambda_3(\lambda_2-\lambda_3)e-\lambda, t]$$

I

Numerical Methods

Example 44

The Highway Code gives the values of the distance D travelled by a car before coming to rest from a speed V

V(miles/h)	20	30	40	50	60
D(ft)	45	75	120	175	240

Find a relationship of the form $D = a + bV + cV^2$ which represents the given data. Comment upon the validity of the expression for speeds below 20 miles/h. Suggest a simple relation between D and V which could be applied for such speeds. (CEI)

In order to scale the equations to avoid high values in V^3, V^4 we first look for a relationship

$$D = \alpha + \beta W + \gamma W^2$$

where $W = V/10$.
The normal equations for a quadratic fit are

$$\Sigma D_i = 5\alpha + \beta \Sigma W_i + \gamma \Sigma W_i^2$$

$$\Sigma W_i D_i = \alpha \Sigma W_i + \beta \Sigma W_i^2 + \gamma \Sigma W_i^3$$

$$\Sigma W_i^2 D_i = \alpha \Sigma W_i^2 + \beta \Sigma W_i^3 + \gamma \Sigma W_i^4$$

Evaluating the required sums and substituting, we obtain

$$655 = 5\alpha + 20\beta + 90\gamma$$

$$3110 = 20\alpha + 90\beta + 440\gamma$$

$$15\ 790 = 90\alpha + 440\beta + 2774\gamma$$

These have solution $\alpha = -62.821$, $\beta = 47.755$, $\gamma = 0.1556$ and hence

$$D = -62.821 + 47.755\left(\frac{V}{10}\right) + 0.1556\left(\frac{V}{10}\right)^2$$

that is

$$d = -62.821 + 4.7755V + 0.001556V^2$$

Now at $V = 20$ the sizes of the three terms are 62.821, 95.51, 0.6224 respectively. The third term is almost negligible in comparison with the other two and will become more insignificant as V decreases.

Hence a relation $D = a + bV$ would be better since it is simpler to apply.

Example 45

(a) Find, to three decimal places, the first positive root of the equation $\tan u = u$ using basic iteration and the Newton–Raphson method.
(b) The formula $x_{n+1} = 2x_n - Ax_n^2$ is to be used to find the reciprocal of A. Determine the limits on the initial guess, x_0, in order that it should converge to $1/A$. (LUT)

(a) The equation $u = \tan u = F(u)$ has associated scale factor $|F'(u)| = \sec^2 u \geqslant 1$; but we seek a rearrangement which has a scale factor < 1 in the neighbourhood of the root.

We try $u = \tan^{-1} u$; the scale factor is $1/(1 + u^2) < 1$ away from $u = 0$.

To locate the root we use a sketch; Figure 39. Note that $y = u$ and $y = \tan u$ both have slope 1 as they pass through the origin.

From the sketch it is seen that the root we require is just less than $3\pi/2$. We take $u_0 = 4.7$ and apply the iterative formula $u_{n+1} = \tan^{-1} u_n$.

The first three iterations are shown in Table 3.

(Remember, u is in radians)

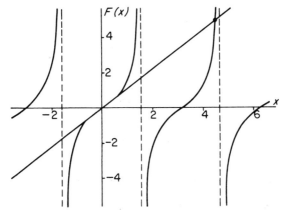

Figure 39

Table 3

n	0	1	2	3
u_n	4.7	1.3612	0.9372	0.7530

In fact, the iterations converge on $u = 0$. The reason lies in the iterative formula. We look for a root near $3\pi/2$, yet $\tan^{-1} u_n$ as provided by calculator or computer must lie in the range $[-\pi/2, \pi/2]$ which puts us on the wrong branch of the tan curve. If we use $u_{n+1} = \tan^{-1} u_n + \pi$ we get the results of Table 4.

The required result is $u = 4.493$.

The Newton–Raphson formula for this equation in the form $\tan u - u = 0$ is

$$u_{n+1} = u_n - \frac{(\tan u_n - u_n)}{\sec^2 u_n - 1} = u_n - \frac{(\tan u_n - u_n)}{\tan^2 u_n}$$

(This latter form requires only u_n and $\tan u_n$.)

Table 4

n	0	1	2	3	4
u_n	4.7	4.5027	4.4938	4.4934	4.4934

94

Table 5

n	u_n
0	4.7
1	4.6883
2	4.6670
3	4.6312
4	4.5805
5	4.5284
6	4.4991
7	4.4936
8	4.4934

The results are shown to 4dp in Table 5.
The relative slowness of the Newton–Raphson method is unusual.

(b) Note that if $x_{n+1} = x_n$,

$$x_n = 2x_n - Ax_n^2 \quad \text{i.e.} \quad x_n = 1/A$$

so that the formula will converge to the correct result. Let

$$x_n = \frac{1}{A} + \varepsilon_n, \quad x_{n+1} = \frac{1}{A} + \varepsilon_{n+1}.$$

On substitution we find that

$$\varepsilon_{n+1} = -A\varepsilon_n^2 = (-A\varepsilon_n).\varepsilon_n.$$

(Note that, had we ignored ε_n^2 we should have had a useless result.)

For the scale factor to be less than 1 in magnitude

$$|A\varepsilon_n| < 1 \quad \text{or} \quad \left| A\left(x_n - \frac{1}{A}\right) \right| < 1.$$

Assume $A > 0$, then

$$-\frac{1}{A} < x_n - \frac{1}{A} < \frac{1}{A}$$

i.e.

$$0 < x_n < \frac{2}{A}.$$

Once a guess x_n is in this range (symmetrical about $1/A$) then all subsequent guesses will be (why?)

Hence

$$0 < x_0 < \frac{2}{A} \qquad (i)$$

A reasonable objection is that since we are trying to estimate $1/A$ we do not know $2/A$; say we want to find $1/81$ then instead of $0 < x_0 < (2/81)$ we insist on $0 < x_0 < (2/100)$. Note that had we taken the equation associated with the formula, namely $x = 2x - Ax^2 = F(x)$ then the criterion $|F'(x)| < 1$ leads to

$$\frac{1}{2A} < x < \frac{3}{2A} \qquad (ii)$$

There is no incompatibility between these results.
(i) says that outside the interval $(0, 2/A)$ the formula *will not* converge.
(ii) says that outside the narrower interval $(1/2A, 3/2A)$ the formula *may not* converge.
The former approach, being tailormade for the specific equation, is more precise.

Example 46

Determine numerically

$$I = \int_0^{\pi/4} \sin x \, \mathrm{d}x$$

by using Simpson's rule with a step length so chosen that the truncation error is less than 0.001. Find a bound for the round-off error in your answer.   (CEI)

The maximum truncation error in Simpson's rule applied to $\int_a^b f(x) \, \mathrm{d}x$ with n strips of width h is given by

$$|\varepsilon_s| \leqslant \frac{(b-a)h^4}{180} M_s$$

where

$$M_s = \max_{[a,b]} |f^{iv}(x)|.$$

In this instance $f^{iv}(x) = \sin x$ and in $[0, \pi/4]$ $\sin x$ is increasing,

96

therefore

$$M_s = |\sin \pi/4| = \frac{1}{\sqrt{2}}.$$

Hence

$$\frac{(\pi/4 - 0)h^4}{180} \frac{1}{\sqrt{2}} < 0.001$$

and since $nh = \pi/4 - 0$ we require

$$n^4 > \left(\frac{\pi}{4}\right)^5 \frac{1000}{180\sqrt{2}}$$

from which it follows that

$$n > 1.174 \text{ (3dp)}.$$

We therefore choose $n = 2$, and work to 5dp. Table 6 summarizes the calculation.
Therefore

$$I_s = \frac{h}{3} \Sigma \text{ terms} = \frac{1}{3} \cdot \frac{\pi}{8} \cdot 2.237\,84$$

$$= 0.292\,93$$

(Note that the value of the integral is $[\cos x]_{\pi/4}^0 = 0.292\,89$ (5dp))

If the value of the integral were not obtainable analytically we should have to quote our result as $I = 0.293 \pm 0.001$.
In our calculation of Σ terms we are effectively adding together 5 numbers (the other is 0), each of which has a maximum round-off error of $\pm 0.000\,005$ so that the sum will have a value with maximum round-off error of $\pm 0.000\,025$. Having used a calculator to evaluate $(\pi/24)$,

Table 6

x	$\sin x$	Coefficient	Term
0	0	1	0
$\pi/8$	0.382\,68	4	1.530\,73
$\pi/4$	0.707\,11	1	0.707\,11
		Σ	2.237\,84

we may assume negligible round-off error. Now since the value of $(\pi/24)$ is approximately between $\frac{1}{7}$ and $\frac{1}{8}$, the final maximum round-off error is about $\pm 0.000\,004$ which is much less than the assumed maximum truncation error and can be ignored.

Example 47

The response $y(>0)$ of a certain hydraulic valve subject to sinusoidal input variation is given by

$$\frac{dy}{dt} = \sqrt{2\left(1 - \frac{y^2}{\sin^2 t}\right)} \qquad \text{with } y_0 = 0, \ t_0 = 0.$$

Why is the Taylor series method unsuitable as a method of solution? Show that

$$\left(\frac{dy}{dt}\right)_0 = \sqrt{\frac{2}{3}}$$

and hence use the Runge–Kutta (fourth order) method to obtain a solution at $t = 0.2$. (CEI)

For the use of Taylor (Maclaurin) series, it is necessary to have the values of y, $\dot{y}$, $\ddot{y}$, etc. at $t = 0$.

It is not easy to obtain $\dot{y}$ at $t = 0$, necessitating the use of L'Hôpital's rule (see Problem 22(b)). To obtain a formula for $\ddot{y}$ would be very tedious and it would be impracticable to try for higher derivatives.

We shall illustrate the use of the Runge–Kutta method by taking $h = 0.2$ and using one step.

Here,

$$\frac{dy}{dt} = f(t, y) = \sqrt{2\left(1 - \frac{y^2}{\sin^2 t}\right)}.$$

$$k_1 = h \cdot f(0, 0) = 0.2\left[\sqrt{\frac{2}{3}}\right] = 0.163\,30 \qquad \text{(5dp)}$$

$$k_2 = h \cdot f(0.1, 0.081\,65) = 0.2\left[\sqrt{2\left(1 - \left[\frac{0.081\,65}{\sin(0.1)}\right]^2\right)}\right] = 0.162\,75 \quad \text{(5dp)}$$

(remember, angles are in radians)

$$k_3 = h \cdot f(0.1, 0.086\,22) = 0.2\left[\sqrt{2\left(1 - \left[\frac{0.086\,22}{\sin(0.1)}\right]^2\right)}\right] = 0.163\,85 \quad \text{(5dp)}$$

$$k_4 = h \cdot f(0.2, 0.163\,85) = 0.2\left[\sqrt{2\left(1 - \left[\frac{0.163\,85}{\sin(0.2)}\right]^2\right)}\right] = 0.159\,95 \quad \text{(5dp)}$$

therefore

$$y(0.2) = y(0) + \tfrac{1}{6}(k_1 + 2k_2 + 2k_3 + k_4)$$

$$= \underline{0.1626} \quad \text{(4dp)}$$

Repeat the process from $t = 0$ using two steps of width $h = 0.1$ and compare your result.

Example 48

Van der Pol's equation, which arises in electronics, is

$$\ddot{x} + (1 - x^2)\dot{x} + x = 0$$

With initial conditions $x = 0.5$, $\dot{x} = 0$ at $t = 0$, solve for x, $\dot{x}$ and $\ddot{x}$ at $t = 0.1$, 0.2. (LUT)

The equation is split into two first-order differential equations

$$\dot{x} = v, \qquad \dot{v} = (x^2 - 1)v - x = g(x, v).$$

The Runge–Kutta fourth-order method is applied as follows:

$$x_{n+1} = x_n + \tfrac{1}{6}(k_1 + 2k_2 + 2k_3 + k_4)$$

$$v_{n+1} = v_n + \tfrac{1}{6}(l_1 + 2l_2 + 2l_3 + l_4)$$

where

$$k_1 = h \cdot v_n, \qquad\qquad l_1 = h \cdot g(x_n, v_n)$$
$$k_2 = h \cdot (v_n + \tfrac{1}{2}l_1), \qquad l_2 = h \cdot g(x_n + \tfrac{1}{2}k_1, v_n + \tfrac{1}{2}l_1)$$
$$k_3 = h \cdot (v_n + \tfrac{1}{2}l_2), \qquad l_3 = h \cdot g(x_n + \tfrac{1}{2}k_2, v_n + \tfrac{1}{2}l_2)$$
$$k_4 = h(v_n + l_3), \qquad\quad l_4 = hg(x_n + k_3, v_n + l_3)$$

and

$$h = t_{n+1} - t_n$$

In our example, $t_0 = 0$, $h = 0.1$, $x_0 = 0.5$, $v_0 = 0$.
Therefore

$k_1 = 0.1(0) = 0,$ $\qquad\qquad$ $l_1 = 0.1 . g(0.5, 0) = -0.05$

$k_2 = 0.1(-0.025) = -0.002\ 5,$ $\quad l_2 = 0.1 . g(0.5, -0.025)$
$\qquad\qquad\qquad\qquad\qquad\qquad = -0.048\ 125$

$k_3 = 0.1(-0.024\ 062\ 5)$ $\qquad\qquad l_3 = 0.1 . g(0.49975, -0.902\ 406\ 25)$
$\quad = -0.002\ 406\ 25,$ $\qquad\qquad\qquad = -0.048\ 169\ 7$

$k_4 = 0.1(-0.048\ 169\ 7)$ $\qquad\qquad l_3 = 0.1 . g(0.499\ 759\ 4, -0.004\ 816\ 97)$
$\quad = -0.044\ 816\ 97,$ $\qquad\qquad\qquad = -0.049\ 614\ 6$

therefore

$$\underline{x_1 = 0.4976 \quad \text{(4dp)}}, \quad \underline{v_1 = -0.0487 \quad \text{(4dp)}}$$

and from the differential equation,

$$\ddot{x}_1 = (x_1^2 - 1)v_1 - x_1 = \underline{-0.4610} \quad \text{(4dp)}.$$

With these starting values we advance to $t = 0.2$. You can check the values below.

$$k_1 = -0.004\ 87, \qquad l_1 = -0.046\ 096$$

$$k_2 = -0.007\ 175, \qquad l_2 = -0.044\ 108$$

$$k_3 = -0.007\ 045, \qquad l_3 = -0.044\ 245$$

$$k_4 = -0.009\ 294, \qquad l_4 = -0.042\ 574$$

$$\underline{x_2 = 0.491\ 2} \quad \text{(4dp)} \qquad \underline{v_2 = -0.092\ 9} \quad \text{(4dp)}$$

$$\underline{\ddot{x}_2 = -0.420\ 7} \quad \text{(4dp)}$$

J
Statistics and Probability

Example 49

(a) In a bolt factory machines A, B, C manufacture 25, 35, and 40 per cent respectively of the total output. Of their outputs, 5, 4, and 2 per cent respectively are defective bolts. A bolt is drawn at random and found to be defective. What are the probabilities that it was manufactured by machines A, B, or C respectively?

(b) A box contains four bad tubes and six good tubes.

If the tubes are checked by drawing a tube at random, testing it and repeating the process (not replacing the tube tested) until all four bad tubes are located, what is the probability that the fourth bad tube will be found (i) at the fifth test (ii) at the tenth test?

(LUT)

(a) This problem demonstrates that probability questions are often most easily tackled from first principles.

Consider an output of 10 000 bolts.

	A	B	C	Total
Number produced	2500	3500	4000	10 000
Defective items	125	140	80	345

Given that a bolt is defective, it must be among the 345; hence the probability that it came from A is 125/345, from B is 140/345, and from C is 80/345.

(b) (i) Let B stand for bad, G for good and x for either. The outcome of the five test results is generally expressed as

$$x \ x \ x \ x \ B$$

where the $4x$ comprise $3B$ and $1G$.
One possible outcome is $BBBGB$

with probability $\dfrac{4}{10} \cdot \dfrac{3}{9} \cdot \dfrac{2}{8} \cdot \dfrac{6}{7} \times \dfrac{1}{6} = \dfrac{1}{35} \times \dfrac{1}{6}$.

It is easy to show that the three other outcomes also have this probability. Then the overall probability is

$$4 \times \dfrac{1}{35} \times \dfrac{1}{6} = \dfrac{2}{\underline{105}}.$$

(ii) A similar argument to that in (a) gives the overall probability as

$$^9C_3 \times \left(\dfrac{4}{10} \times \dfrac{3}{9} \times \dfrac{2}{8} \times \dfrac{6}{7} \times \dfrac{5}{6} \times \dfrac{4}{5} \times \dfrac{3}{4} \times \dfrac{2}{3} \times \dfrac{1}{2} \right) \times 1$$

the last 1 being the result of specifying the outcomes of the first nine tests which leaves the outcome of the tenth test a certainty. The overall probability reduces to 2/5.
This is such a simple result that it should be possible to determine it more easily.
Convince yourself that the problem is equivalent to finding the probability of leaving one bad tube on one side, to be picked up at test 10. This situation has probability $\frac{4}{10} = \frac{2}{5}$.

Example 50

(a) It is known that 10 per cent of the output of a certain manufacturing process is unsatisfactory and considered defective.
Find the probability that out of a large batch
(i) a sample of 20 contains exactly two defective items,
(ii) a sample of 10 contains not more than one defective item.
(b) In a survey of a day's production from 400 machines making similar components, four items selected at random from the output of each machine were inspected in detail. The number m of

machines producing f faulty items was found to be

f	0	1	2	3	4
m	16	89	145	118	32

Represent this data by a binomial distribution and calculate the theoretical distribution of faulty components.　　　　　(EC)

(a)　The fact that the batch is large allows the use of the binomial model; in particular, we can assume that the probability of a single item selected being defective is 0.1.

(i)　We apply the binomial model with $n = 20$, $p = 0.1$

$$p(2) = {}^{20}C_2(0.1)^2(0.9)^{18} = 0.2852.$$

(ii)　This time, the binomial model with $n = 10$, $p = 0.1$ is used.

$$p(0) = (0.9)^{10} = 0.3487; \quad p(1) = 10(0.1)(0.9)^9 = 0.3874.$$

Then $p(\leqslant 1) = p(0) + p(1) = 0.7361$.

(b)　The first problem here is that of non-standard notation: the frequency (usually f) is m and the value of the random variable (usually x) is f. If we are to use a binomial model we need to know n, the number of trials, and p, the probability of success on any one trial.

$$\sum f_i m_i = 861; \qquad \bar{f} = 861/400 = 2.15 \qquad (3\text{sf})$$

The expected value for the binomial distribution is np. In this example $np = 4p$ and is matched with the experimental value $\bar{f}$. Hence $p \simeq 0.54$.

An example calculation shows that in this model the probability of two faulty items in the batch of four tested is

$$p(2) = {}^4C_2(0.54)^2(0.46)^2.$$

The frequency $400p(2) \simeq 148.3$ is compared with the experimental frequency 145. Table 7 below shows the full comparison.

Table 7

f	0	1	2	3	4
m (Experimental)	16	89	145	118	32
m (Theoretical)	18.2	84.8	148.3	115.2	33.5

Example 51

(a) The number of days n in a 100-day period when x accidents occurred in a factory is shown.

No. of accidents (x)	0	1	2	3	4	5
No. of days (n)	42	35	14	6	2	1

  (i) Fit a Poisson distribution to this data to predict the probability of x accidents occurring.
  (ii) Compare the variance of the given and calculated distributions. (CEI)

(b) A machine-shop storekeeper finds that, over a long period, the average demand per week for a certain machine tool is 3. His stocking policy is to make up stock to 4 at the beginning of each week. Estimate the probability that he will fail to satisfy demand in a given week, and determine his stocking policy if the chance of running out is not to exceed 5 per cent. (LUT)

(a) If we use a Poisson model we need only decide on the value of λ; we choose this to agree with the mean of the given data.
  (i) $\lambda = (\sum n_i x_i)/100 = 0.94 = \sigma^2$, since for Poisson, mean = variance.
  The Poisson probabilities are calculated from the formula

  $$p(r) = \frac{\lambda^r}{r!} e^{-\lambda} \quad \text{or via} \quad p(0) = e^{-\lambda}, \qquad p(r) = \frac{\lambda}{r} p(r-1).$$

No. of accidents	0	1	2	3	4	5
Probability of occurrence	0.3906	0.3672	0.1726	0.0541	0.0127	0.0024
Predicted frequency	39.06	36.72	17.26	5.41	1.27	0.24
Actual frequency	42	35	14	6	2	1

  (The theoretical probabilities add to 0.9996, due to round-off.)
  (ii) From the frequency table provided we calculate $s^2 = 1.15$. This is fairly close to the theoretical value of 0.94.

(b) The sensible unit of time to take is a week. Hence we need to find λ in units of number of demands/week.
  The expected value for the Poisson model is $\lambda = 3$.

We calculate $p(0)$, $p(1)$, $p(2)$, $p(3)$ and $p(4)$.

The probability of failing to satisfy demand in a week is

$$1 - p(0) - p(1) - p(2) - p(3) - p(4) = \underline{0.185}.$$

Let the correct policy be to stock n items. We then require that there are up to n demands in a week with probability of at least 0.95, i.e.

$$p(0) + p(1) + p(2) + \ldots + p(n) > 0.95$$

whereas

$$p(0) + p(1) + p(2) + \ldots + p(n-1) \leqslant 0.95.$$

In practice we continue to calculate Poisson probabilities and accumulate them until the running total first exceeds 0.95. This occurs when $\underline{n = 6}$.

Example 52

(a) In a manufacturing process, a piston of circular cross-section has to fit into a similarly shaped cylinder. The distributions of diameters of pistons and cylinders are normal with parameters:

Pistons: mean diameter 10.42 cm, standard deviation 0.03 cm
Cylinders: mean diameter 10.52 cm, standard deviation 0.04 cm.

If the pistons and cylinders are selected at random for assembly
  (i) What proportion of the pistons will not fit into cylinders?
  (ii) What is the chance that, in 100 pairs selected at random, all the pistons will fit?
                                                              (CEI)

(b) Wire cables are formed from ten separate wires, the strength of each wire being normally distributed with standard deviation of 24.5 N about a mean of 645 N. Assuming that the strength of each cable is the combined strength of the separate wires, what proportion of the cables have a breaking strain of less than 6350 N? Assuming the variability to remain unchanged, to what must the mean strength of the individual wires be increased if only 1 in 1000 cables is to have a breaking strain of less than 6350 N?
                                                              (LUT)

(a) (i) The distribution of the difference (cylinder diameter − piston diameter) is

$$N([0.04 - 0.03], [(0.04)^2 + (0.03)^2]) = N(0.1, (0.05)^2)$$

A pair will fail to fit if this difference is less than zero. To use the standard normal distribution we need

$$z = (x - \mu)/\sigma$$

We calculate $z = (0 - 0.1)/0.05 = -2$ which yields a probability of not fitting of 0.0228 or 2.28%. See Figure 40(a).

(ii) Now we use a binomial model with $n = 100$,

$$p = 0.0228 \quad \text{and} \quad q = 0.9772.$$

The probability that all 100 pairs will fit is

$$(0.9772)^{100} = 0.0996.$$

(b) The distribution of cable strengths is $N(6450, [\sqrt{10}(24.5)]^2)$. The appropriate calculation is

$$z = \frac{6350 - 6450}{\sqrt{10}(24.5)} = -1.291$$

and this produces a probability that the breaking strain of a cable will be < 6350 of 0.0983. See Figure 40(b).

If we want a probability of 0.001 then we need $z = -3.1$. This leads back to a mean breaking strain of $6350 + 3.1\sqrt{10}(24.5) = 6590$ N

Hence the m.b.s. of a wire is 659 N.

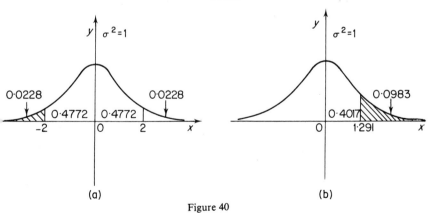

(a)   (b)

Figure 40

Example 53

Routine strength tests have established that certain spars have a breaking stress normally distributed about 27.56 N/m² with a standard deviation of 1.01 N/m².

(a) What is the probability that the average breaking stress from a sample of 10 spars will be less than 27.24 N/m²?

(b) Certain modifications are made in the construction of the spars and in order to test whether these modifications have changed the mean strength, a sample of 10 modified spars was tested. They gave a sample mean of 28.53 N/m². Does this suggest that the modifications have had a real effect?   (LUT)

(a) The distribution of breaking stress is $N(27.56, (1.01)^2)$. We first calculate

$$z = \frac{27.24 - 27.56}{1.01/\sqrt{10}} = -1.00.$$

The appropriate probability is therefore 0.1587.

(b) If we decide that the purpose of modification is to *increase* the mean breaking stress then we employ a one-tail test (see Figure 41(a)). At 5 per cent level of significance the rejection region is $z > 1.64$; at 1 per cent level of significance it is $z > 2.33$. If, however, the modifications are investigated to see whether they have *changed* the m.b.s. then we use a two-tail test (see Figure 41(b), (c)). At 5 per cent level of significance the rejection region is $|z| > 1.96$; at 1 per cent level of significance it is $|z| > 2.57$.

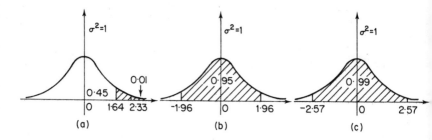

Figure 41

We calculate

$$z = \frac{28.53 - 27.56}{1.01/\sqrt{10}} = 3.04.$$

Under all circumstances above we conclude that the modifications *have* been effective.

Note that if a two-tail test is employed we can proceed alternatively using the idea of a confidence interval. For example, with 95 per cent confidence we can say that the mean breaking stress lies in the interval

$$28.53 \pm 1.96\left(\frac{1.01}{\sqrt{10}}\right) = \underline{(27.90, 29.16)}$$

and this does *not* include the previous m.b.s. value of 27.24. If we decide that the modifications *have* altered the m.b.s. then our point estimate is now 28.53.

SOME BASIC RESULTS

Section A

(i) If $z = x + iy$ and $\bar{z} = x - iy$ then $z\bar{z} = |z|^2$; $z\bar{z}$ and $z + \bar{z}$ are both real.

(ii) $e^{i(\theta + 2k\pi)} \equiv e^{i\theta}$ for $k = 0, \pm 1, \pm 2, \ldots$

Section B

If $\mathbf{a} = (a_1, a_2, a_3)$, $\mathbf{b} = (b_1, b_2, b_3)$, $\mathbf{c} = (c_1, c_2, c_3)$

then $\mathbf{a} \cdot \mathbf{b} = a_1 b_1 + a_2 b_2 + a_3 b_3$;

$$\mathbf{a} \wedge \mathbf{b} = (a_2 b_3 - a_3 b_2, a_3 b_1 - a_1 b_3, a_1 b_2 - a_2 b_1);$$

$$\mathbf{a} \wedge \mathbf{b} \cdot \mathbf{c} = \mathbf{a} \cdot \mathbf{b} \wedge \mathbf{c} = \begin{vmatrix} a_1 & a_2 & a_3 \\ b_1 & b_2 & b_3 \\ c_1 & c_2 & c_3 \end{vmatrix}$$

Section E

The Maclaurin's series for a function $f(x)$ is

$$f(x) = f(0) + xf'(0) + \frac{x^2}{2!} f''(0) + \ldots + \frac{x^r}{r!} f^{(r)}(0) + \ldots$$

Section F

(i)   If $u = u(x, y)$ then the total differential

$$\mathrm{d}u = \frac{\partial u}{\partial x}\,\mathrm{d}x + \frac{\partial u}{\partial y}\,\mathrm{d}y.$$

(ii)   If $x = x(t)$, $y = y(t)$ then

$$\frac{\mathrm{d}u}{\mathrm{d}t} = \frac{\partial u}{\partial x}\frac{\mathrm{d}x}{\mathrm{d}t} + \frac{\partial u}{\partial y}\frac{\mathrm{d}y}{\mathrm{d}t}.$$

(iii)   If $f(x, y) = 0$ then $\dfrac{\partial f}{\partial x} + \dfrac{\partial f}{\partial y}\dfrac{\mathrm{d}y}{\mathrm{d}x} = 0.$

Section H

(i)   The first order o.d.e. $\dfrac{\mathrm{d}y}{\mathrm{d}x} + P(x)y = Q(x)$

can be solved by writing $I(x) = \exp\{\int P(x)\,\mathrm{d}x\}$; then

$$\frac{\mathrm{d}}{\mathrm{d}x}\{y.I(x)\} = Q(x).I(x).$$

(ii)   If the Laplace transform of $y(t)$ is $Y(s)$ then the LT of $\mathrm{d}y/\mathrm{d}t$ is $s\,Y(s) - y(0)$ and the LT of $\mathrm{d}^2y/\mathrm{d}t^2$ is $s^2\,Y(s) - s.y(0) - \dot{y}(0)$.

Section I

(i)   The Newton–Raphson formula for the iterative solution of $f(x) = 0$ is

$$x_{n+1} = x_n - f(x_n)/f'(x_n).$$

(ii)   The Simpson integral approximation to $I = \int_a^{a+2h} f(x)\,\mathrm{d}x$ using two strips is

$$I_s = \frac{h}{3}\,[f(a) + 4.f(a+h) + f(a+2h)].$$

(iii)   The Runge–Kutta fourth order approximation to the solution of the o.d.e. $\mathrm{d}y/\mathrm{d}x = f(x, y)$; $x = x_n$, $y = y_n$ is

$$x_{n+1} = x_n + h$$
$$y_{n+1} = y_n + \tfrac{1}{6}(k_1 + 2k_2 + 2k_3 + k_4)$$

where

$$k_1 = h \cdot f(x_n, y_n)$$
$$k_2 = h \cdot f(x_n + \tfrac{1}{2}h, y_n + \tfrac{1}{2}k_1)$$
$$k_3 = h \cdot f(x_n + \tfrac{1}{2}h, y_n + \tfrac{1}{2}k_2)$$
$$k_4 = h \cdot f(x_n + h, y_n + k_3).$$

Section J

(i) The binomial probability model gives the probability of r successes in n independent trials as

$$p(r) = {}^nC_r p^r (1 - p)^{n-r}$$

where p is the probability of success on one trial.

(ii) If x_1 follows the distribution $N(\mu_1, \sigma_1^2)$ and x_2 follows the distribution $N(\mu_2, \sigma_2^2)$ and x_1 and x_2 are independent random variables then $(x_1 + x_2)$ follows $N(\mu_1 + \mu_2, \sigma_1^2 + \sigma_2^2)$ and $(x_1 - x_2)$ follows $N(\mu_1 - \mu_2, \sigma_1^2 + \sigma_2^2)$.

(iii) If $x_1, x_2, \ldots, x_n$ are independent random variables which follow $N(\mu, \sigma^2)$ then $\bar{x} = (x_1 + x_2 + \ldots + x_n)/n$ follows $N(\mu, \sigma^2/n)$.

Index